Amish Barns Across America

by John M. Zielinski

PENNSYLVANIA
NEW YORK
DELAWARE
MARYLAND
VIRGINIA
WEST VIRGINIA
KENTUCKY
TENNESSEE
FLORIDA
ARKANSAS
TEXAS
OKLAHOMA
MISSOURI
KANSAS
NEBRASKA
MONTANA
MINNESOTA
WISCONSIN
MICHIGAN
OHIO
INDIANA
ILLINOIS
IOWA

ONTARIO

AMISH HERITAGE PUBLICATIONS
Box 2660
Iowa City, IA 52244
(319) 683-2714

THE AMISH ACROSS AMERICA 136 pages 8 1/2 X 11		$12.95 paperback
Contains nearly 200 photographs in color and black and white		$17.95 hardcover
AMISH CHILDREN ACROSS AMERICA	ISBN No. 0-910381-20-8	$5.95 booklet
AMISH HORSEFARMING ACROSS AMERICA	ISBN No. 0-910381-17-8	$4.95 booklet
AMISH BARNS ACROSS AMERICA 144 pages, 16 in color	ISBN No. 0-910381-13-5	$10.95 paperback
AMISH LIFE THROUGH A CHILD'S EYES by Alma Hershberger		$8.95 hardback

An intimate look at Amish life thru a little girl's eyes as she grew up near Olwein, Iowa

ART OF AMISH COOKING by Alma Hershberger $9.95 softbound
    Over 650 new recipes calling on the Yoders, Beachy, Chupps, et al.
Please add $1.25 for 1st book, .75 each additional, for shipping

<u>FORTHCOMING BOOKS IN THE SERIES:</u>
AMISH WOMEN ACROSS AMERICA , with Alma Hershberger
    Born and raised Amish - presents a woman's viewpoint. Planned for 64 pages $5.95 tentative
ALMA'S AMISH KITCHEN - Book and video tape
    Recipes and reminiscences gleaned from Amish neighbors and relatives across America
OHIO'S AMISH HERITAGE
INDIANA'S AMISH HEARTLAND
OTHER MIDWESTERN BARN BOOKS

THE AMISH HERITAGE BOOK CLUB
Entitles you to purchase books by John M. Zielinski and other authors at a special discount rate for members only. A one time membership fee of ten dollars can give you a 25% discount on all books by John M. Zielinski purchased within 30 days after publication, and 10% off hereafter on all books. This also includes books by other authors which Mr. Zielinski has read and now would like to make available to you. Many of these titles will not be available in bookstores. For years, Mr. Zielinski has explored every publication available on the Amish, searching out obscure titles in the Goshen College Library and new publications in Amish areas across the country. Now you may purchase books from the Amish Heritage Book Club at this special rate. Make checks payable to:AMISH HERITAGE BOOK CLUB
        P.O. Box 2660
        Iowa City, Iowa 52244

Dedication: To Zane Zielinski, my son. I am proud of your attending the Governor's Institute at the U of I. You have shown a special talent as a writer and a special human being and your father is proud. My only hope is that you will have a future in a country without copyright rights.
Special thanks: To my friends across this country who have put me up and listened to me and helped me, especially Leonard and Irene Gross, Marjorie Mentzer, and Alma Hershberger. Thanks also to all the anonymous Amishmen who helped me explore their lives and their barns over the last 25 years
Book Design: Special thanks to Jill Butler Smith, both as book designer and friend. Without your talent, your listening ear, and your patience this book would not be half so well done.

All rights reserved. No part of this publication may be reproduced, stored in a retrieval system, or transmitted, in any form or by any means, electronic, mechanical, photocopying, recording, or otherwise, without prior written permission of the copyright owner.

Copyright 1989 by John M. Zielinski, as Amish Heritage Publications
Printed in the United States of America
ISBN 0-910381-13-5
10 9 8 7 6 5 4 3 2 1

Table of Contents

Preface	5
Birth of the Barn	15
Barn Raising	32
Josie Miller	36
Big Dave Bontrager	41
The Barn as a Church	44
Barn Dances	47
Children and the Barnyard	52
The Barn as Factory	54
Mail Pouch Signs	56
The Horse and the Barn	59
The Barn as an Icon	63
The Weather	66
Round Barns	77
The Amish Farmstead	80
Swiss Barns	83
Barns of Pennsylvania	89
Barns of Ohio	103
Barn wood	112
Barns of Maryland and Delaware	115
Menno-hof Museum	118
Brown Swiss Barn	121
Barns of Indiana	122
Barns of Illinois	129
Barns of Missouri	132
Barns of Kansas	133
Barns of Iowa	135
A Typical Day	141
Amish Area Zip Codes	143
Grandpa's Barn	144

This barn is from an area near Berne, Switzerland, on a farm still occupied by descendents of the Amish/Mennonites. This barn shed is over 300 years old. The cattle barn is part of the house.

Preserve and Restore

Throughout America old barns are taking on a new look, leaking roofs are being fixed, worn siding and floorboards are replaced, cupolas are being repaired, and a fresh coat of paint is applied. All this is being done not as some vast nationwide effort to restore the American barn, but because of the tremendous need of the old Order Amish to find new land for their expanding families. The Amish now are over 100,000 strong with the population expected to double within the next twenty years. Far from disappearing, the Amish are becoming a common sight in farmlands across America.

Preface

The title of this book is actually a misnomer since many of the barns in this book may not have been built by the Amish, but today they are Amish barns because the Amish are moving into farm areas all over America. Barns that have been neglected for decades are starting to take on a new life as Amish farmers put them back to their original duties. Old fashioned barns built with oak beams and wooden pins are being built today across America as the modern farms of 1,000 or more acres are being broken up once again to farms of 80 to 160 acres.

New pioneer Amish families are having to live in the way of the pioneers of old. Near Kidron, Ohio an Amishman divided a larger farm in two, built a large barn and housed his family in one corner of the newly built barn for a year until he could get the new house built. There is an old saying: "A good barn built many a fine house, but no fine houses ever built good barns." The barn came first. The family lived in a lean-to or log cabin until the barn helped the family prosper.

This book is a celebration of the American barn and the role that it played in the growth of this country. What we tend to think of as in the past, still is and will continue to be in Amish life. The barn still exists for them, as farm factory, as social center where church is held--with funerals, marriages and baptisms. It is the center for fun; for children at play, and a place to hold barn dances for teenagers. This book is about what the barn was yesterday to all American farm families, and what it remains today for the Amish. It is my hope that through their eyes and my camera lens you will be able to relive the building of farmland America and the creation of that unique America structure--the BARN.

The Amish farm is seen as old fashioned by many, but it is perhaps the farm of the future, as Amish families and new Amish farms are spreading across farmland America. Eleven new Amish settlements have been formed in New York state since 1974, more Amish families are moving into Michigan, Wisconsin, Minnesota, and Missouri. Since the first Amish settlement in Missouri in 1953, a whole string of new settlements have been founded and finally have spilled over into Arkansas. As a very prosperous and dynamic people, the Amish seek more and more farmland for their exploding population. Since 1960, 123 new Amish settlements have been founded in rural America. That means that seventy percent of the total number of Amish settlements have been founded since 1960.

The average number of Amish children per family is 7.1, which is nearly four times the national average. What does this mean to the barns in America? Let an Amish farmer shopping for a new settlement in New York explain: "We look for an area where we can see that it was once prosperous. The barns around here (New York State) are for the most part large--100 to 125 feet long. You know those farms had to be producing well to support those barns. If they did once, they can again."

Each Amish group that moves into a new area brings with it connections that would allow the group to bring hundreds of relatives, friends and fellow Amish from nearby states to rebuild and repair older barns or to build new ones if necessary. If one examines the march of Amish farm communities across New York, it is evident that they may soon be sitting on New York City's doorstep.

This has not escaped the notice of the *New York Times* and the *Wall Street Journal.* In 1988, they cited the Amish as living proof that the get-big-or-get-out theory that dominated American agricultural thinking for the last 50 years was no longer unchallenged. The Times stated that an Amish farm could produce $125,000 in gross annual income from 80 acres. Only $6,000 to $8,000 is needed to provide food, clothing, and shelter for a family of six or more children, and a number of grandparents.

It seems the Amish family farms are proving themselves successful, while the modern farms demand an increasing supply of chemical fertilizers and energy to produce the same amount of crop. From 1980 to 1985, the amount spent by American agriculture on fertilizer, weed killer and pesticides grew from $3.5 to nearly $4.4 billion dollars. This does not take into account the millions of dollars in fuel necessary to spread these harmful and

Lancaster, Pa. area. These children need lots of clean clothes. See the clothesline.

potentially dangerous and life threatening chemicals on the land.

Contrast this with the Amish farmer, who will use insecticides and fertilizer when necessary, especially when taking over a new farm in such places as New York, but will gradually phase out these items in favor of more natural methods. Amishman David Kline writing in *National Wildlife* told how he encourages cliff swallows to nest along the side of his barn, because each nesting pair collect some 900 insects a day to feed their young. Starting with six nests the first year, Kline aided their spread by adding wooden cleats under the eaves of his barn and now has more than 200 nesting pairs. The birds consume some 180,000 insects per day. Composting, manuring, crop rotation, and other conservation techniques help enrich the soil and keep it productive to a far higher degree than with modern chemical methods.

Far from being a changeless people, the Amish of the past were in the forefront of the agricultural revolution. They are credited with introducing the seeding of clover into their areas to improve the productivity of the soil and to yield more feed for livestock, which in turn yielded more manure to help build better soil.

As the yields improved, the Amish harvested hay and clover and began stall feeding their animals,. Stall feeding in turn led to bigger and better barns for the storage of feed and grains, and as housing for multiplying livestock.

Thomas Jefferson might well have been speaking of the Amish when he wrote: "Those who labor on the earth are the chosen people of God."

Jefferson saw the farmer as the most "independent and virtuous of citizens." Viewing the Amish farmer today, he is still independent and virtuous, free from farm subsidies, and other government sponsored programs.

With little more than today's equivalent of an axe and adze, the Amish can build their own homes, barns, and outbuildings. A recent article from the New Philadelphia, Ohio, newspaper tells of the Amish felling 125 trees to cut up 25,000 board feet of lumber to rebuild a barn. Who today conceives of a barn being built by first going out into the woods and logging 125 selected trees? Yet this was the way it was done in the settling of America.

Near Kalona, Iowa.   Thunderhead!

From the time of Thomas Jefferson to the present the Amish continue to be in the vanguard of pioneers settling new frontiers. They originally settled in Berks and Lancaster counties in the early 1700's where they built a base of beautiful productive farms and literally invented the America barn and the American farmstead. No buildings existed in Europe the size and scope of the American barn and farmstead; hence, the early barns were borrowed from existing castles and other buildings built in the Rhine Valley of Germany and in Switzerland. Some of the oldest standing barns built of stone with oak framework, have what look like gun slits in castle walls Since the Lancaster valley is in some ways as hilly as Switzerland, it was handy to build a banked barn, or one set into a hill so the livestock was housed below and you could drive into the upper story and unload hay and grains which could then be thrown down to feed the livestock. The grains and hay also acted as a insulator against the cold winters, while other insulation was gained by setting it into a hill.

The purpose of this book is not only to celebrate the barn, and the barn heritage of America as embodied in the surviving structures found throughout Amish areas; but to celebrate the rural horsefarming way of life they have kept alive. This means barns are not ghost structures, wasting away beside some long abandoned house, but living breathing embodiments of America's farm past. Today, from places like Minneapolis and Cleveland, Philadelphia and New York, people are drawn into Amish areas to view the farm as it once was all across America. The Amish areas have become living museums.

Writer Jonathan Swift in the early 18th century commented: "Whoever could make two ears of corn or two blades of grass to grow upon a spot of ground where only one grew before would deserve better of mankind and do more essential service in this country than the whole race of politicians put together."

Picking up on this quote, the U.S.'s first commissioner of Agriculture, Isaac Newton in 1863, admonished every young farmer "to make two blades of grass grow where but one grew before."

Sound advice for their day, but today this country and its farmers have started to ask, "two blades at what cost?"

Just east of Intercourse, Pa.

In doubling and doubling again the blades of grass and the ears of corn, farmers have doubled and re-doubled their energy, insecticide, and fertilizer use. What has it gotten the American farmer? Surplus and low prices, and a land in which wells, streams, and rivers are polluted with chemical residue.

According to the Department of Agriculture, almost a third of the nation's 630,000 full-time farmers are in danger of financial collapse. America's heartland is in the midst of what is commonly described as the worst period since the Great Depression. The situation has been described as a financial tornado sweeping across the land, devastating one farm and leaving another untouched.

At the heart of this matter is farm subsidy and an overproduction of corn, soybeans, and other commodities in monumental proportion. The government has had to give away millions of tons of grain, dried milk, cheese, and other products to the needy of this country. Yet should abundant crops bring disaster?

Much of this is occurring because of one-crop farming; farming that requires more and more energy to produce the same amount of corn each year or soybeans or wheat and always more fertilizer, more insecticides more irrigation.

The Amish farmers of Indiana had an obvious leg up on other farmers during the summer drought of 1988. One only needed to drive down country roads with Amish farms on one side and a modern farmer on the other, to notice the Amish corn was a foot higher and seemed less stressed by the drought conditions. The explanation is the Amish method of conservation farming, which uses manure and composting to insure a looser soil and soil filled with worms that help aerate the soil. Field stubble and other loose materials are plowed back in to enrich the soil and encourage it to hold moisture. A morning's dew will add moisture to an Amish corn crop and often because of different planting techniques, evaporates from a modern farmers field.

An old saying might here be refitted: "As the farmer goes, so goes the nation." With over 200,000 farmers holding 100 billion dollars in shaky loans, there is a major debt problem for the nation's banks. And the small towns in the heart of farming America find it harder and harder

Eric Sloane from The Age of Barns comments on the beauty of the plain barn board: "one of nature's special masterpieces--it's composition, grains and shades and knots of weathered gray could be framed and hung just as in a modern painting."

to make ends meet as more and more farmers have less and less to spend.

Contrast this with the Amish farmer who, because of almost explosive population growth, (an average birth rate per family of 7.1), has an increasing farm income thanks to his natural way of farming, use of few chemical fertilizers and insecticides, and the increased yields he can look forward to. The Amish are perpetually in search of new lands for new settlements. Said that same Amishman exploring New York State, "We look for a small town with grain mills. Even if closed, we can reopen them and a growing Amish population can support new business in a old town."

Kalona, Iowa is an example. Twenty-five years ago, when I began my research on the Amish, one of the first things I noticed was that it had three hardware stores, with a population of less than 1400 in town. Yet Iowa City, with a population then in excess of 25,000, had only two hardware stores.

In Iowa and Nebraska at least half the jobs are in some way tied to agriculture. Is it any wonder why small communities are seeking Amish neighbors, for Amish neighbors can spark the rebirth of a town? Within recent years, two states have had a great influx of Amish. Small towns in upper New York State actually came into Amish areas in Pennsylvania and offered land at a good price around their communities. Since 1974, eleven new Amish communities have started in New York to join the earlier one at Conewango Valley, just below Buffalo. Missouri is another state with lots of land and not many farmers left. Now it has been discovered by the Amish and they have begun a steady march from north to south, establishing community after community. The lowest is around Seymour, Missouri, below Bagnell Dam, in the heart of the Ozark recreation country.

Senator Tom Harkin (D-Iowa) recently applauded farm aid concerts as a means to "focus attention on the fact that the old ways of helping farmers are not working." Perhaps a better answer might be to have a long hard look at the old ways of farming.

Harry Bresley of Ord, Nebraska, is a working farmer who uses Percherons to farm. He farms 2,000 acres with his son, Dean, and claims, "I am convinced that any young

Barn raising near Millersburg, Ohio.   Photos by Bruce Glick.

farmer, if he wanted to, could get started on a small amount of money if he would return to horse-drawn equipment." Bresley believes that any man could farm 160 acres (a quarter section) with horses. The Amish most often confine themselves to about 80 acres, preferably with timber, pasture, and cropland that can be rotated between plantings.

Today with both government and industry seeking new ways to alleviate the modern farmers' chemical dependency on fertilizers and insecticides, the surprisingly efficient techniques of the Amish are being studied. From 1980 to 1985, the use of pesticides on farm land increased from three and a half billion dollars to nearly four and a half billion dollars.

How did a group of less than 100,000 men, women, and children become the focus of many government, industry, and university studies? Because they are performing the business of farming with remarkable efficiency. Unlike what many people believe, farm converted over to Amish horsefarming use, although it may be subdivided a number of times, does not decrease in production efficiency but increases. Acres that once produced thirty bushel of oats per acre are now producing sixty and even up to one hundred bushels. This the Amish are doing without the use of more than a minimum of chemical fertilizers and pesticides.

Recent tests of the water quality of Amish farms versus non-Amish farms show much lower amounts of harmful by-products in the water. On <u>modern</u> farms, damage to wells and streams has become so chronic and severe that they pose potential health hazards for nearby communities as well as the farms themselves. The government is being asked to come up with money and new programs to alleviate this poisoning of the land.

Detail of a hand hewn beam from the Johnson/Washington County area, it shows marks of the axe and the adze shown below. The tools are from Eric Sloanes Barn Museum in Connecticut.

The Birth of the Barn

Egypt had its pyramids, Rome its coliseum, medieval Europe its castles and cathedrals; but America has its barns. A cry has been heard throughout America that the barn is disappearing. Those who make such claims have not watched several hundred Amishmen raising a barn of up to one hundred feet in length from a pile of lumber to the finished product in a single day.

Before the Revolutionary War, the Amish were building peg barns in Lancaster, Pennsylvania. They are still doing so today, and in Ohio and Indiana as well. These peg barns' major framework is put together with six by eight inch oak beams locked together with wooden pins. These pins so tightly lock the structure together that such a barn has sometimes been rolled on its side without coming apart. The nailed frame barn began to replace peg barns more than one hundred years ago. It is these framed barn structures that constitute most of the sagging and collapsed barns to be found scattered across the farm belt of mid-America. They might have been easier to build, requiring far less man power, but they were not built with the strength of the Amish barn.

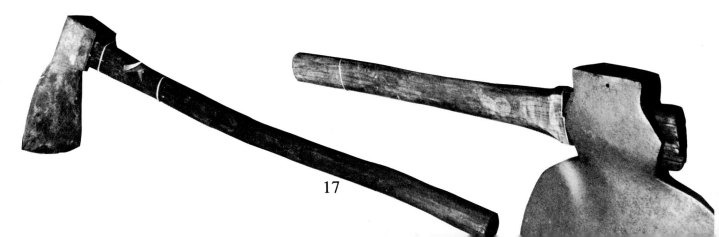

Photographer Joel Troyer, commenting on an Amish barn built three years ago near Sugarcreek, Ohio, said, "Nearly eight hundred men were working on it. They began at seven and by shortly after one o'clock they were putting hay in an almost completed barn. It takes teamwork and the skill passed down for hundreds of years to put together one of these barns. They are built to last and no one hundred years of hard use is going to weaken these structures."

The framing of all the early barns was done in available hardwoods- mostly oak, but walnut and hickory were also used. Usually the floors are sturdy oak planking. The introduction to <u>Pennsylvania German Barns</u>, published by the Pennsylvania German Folklore Society, states that although sawn timbers began to be introduced in the 19th century, pegged barns, whose main structure was formed with the axe and the adze, were still being set up after the Civil War. So exact were the structures that if one barn burned, another barn, no longer in use, could be found, dismantled and re-erected within the fire-scoured, but intact masonry walls of the other.

In reading <u>Pennsylvania German Barns</u>, what becomes apparent is that the American barn was a unique compounded architectural form. From military architecture they took the overhanging framed bay, from walled city and church they took a frame and peg structure made from the ample timbers of Western Europe. John Heyl, in his introduction to <u>Pennsylvania German Barns</u>, said, "The Pennsylvania German in his methods of 'barn raising'... had quickly arrived at an indigenous solution to the farm building two centuries before such Americans as architects Louis Sullivan and Frank Lloyd Wright were demanding...that the form of a building should follow its function or need."

In the Middle Ages, the men who worked the land, whether serf or freeman, lived in walled cities under the shadow of the castle's protection. Each day the men would go forth to till the field and return at night with wagons, tools, and animals to the safety of the city.

When the first settlers came to America they soon found that they could possess land of undreamed size. Each man might be lord of his own manner, so they had

Opposite page: Large detail of beam from the same barn.

The picture in the middle shows tool used to make shake shingles for roofing. The remainder of the tools are from Eric Sloane's Barn Museum. The topmost image show the adze and the broadaxe carving a beam out of a tree trunk.

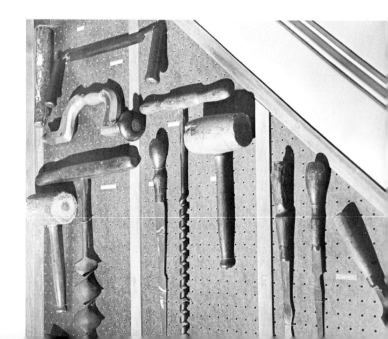

to build their own farmsteads based on individual ownership. Since the majority of the early settlers came from the countryside of Southern Germany, the Rhine Pfalz, and Switzerland, where there were deep valleys, and buildings were often banked into a hill and entered from two levels. The banked barn fit well into the steep, heavily wooded valleys of Pennsylvania.

The Amish who began immigrating to Pennsylvania at the beginning of the 1700's were called the Pennsylvania Dutch in an Americanized version of the word Deutche, meaning "German". They brought with them a skill for farming and building that had made them the most sought after and the most taxed farmers in Europe. The feudal governments of Europe sought them because they worked hard and with great skill, introducing fertilizing and composting, crop rotation, and the planting of clover brought from one area to another. Even the Romans had marveled at the ability of the Germanic people to fashion and build from the heavy hardwood forest of their area of Europe.

"The able use of the axe in pioneer times was the key to comfort in the winters ahead and the fruitfulness of the farm." Heyl wrote. He comments that even the shape of the European axe changed. So handy with an axe and so knowledgeable of the ways of trees were those early Germans that they could notch a row of five or six trees, letting the one chopped down bring down the others. And so it was, that except for those few early stone barns painstakingly put together, sometimes over a period of two years, there evolved a barn so basic that with enough men it could be assembled in a single day.

Earliest of all barns were the simple log structures built by notching the ends in the way of log cabins. At the end of the 18th century, these kinds of barns outnumbered all others, but these were always of modest proportion, capable of being put together by a few men. When the number of neighbors had grown from a dozen to hundreds, it was time then to put up the big barns to match the big farms.

It is interesting to note that the rectangular box structure of the early stone barns in Lancaster - stable on the bottom, banked into the hill, next level threshing floor in the middle with raised areas on either side

Near Gap, Lancaster, County, Pa. On the opposite page see the 1790 written in 1790 in slake lime on the inside of a stone walled barn.

sometimes becoming separate third levels, was copied across America.

The stone was carried over into some structures in eastern Ohio. I discovered two barns from circa 1840. The one barn had a corner stone carefully chiseled out "1838"; the other was occupied by an Amish farmer whose great, great, grandfather had farmed the place. Word was that it was built within two years of the other barn and took two years to construct from sandstone quarried nearby.

The earliest barns are made up of a series of small or medium size squares of logs, suggesting a scarcity of manpower and a need to build in sections. **Josie Miller**, Ohio barn builder, comments on the barns of Holmes and the surrounding counties, saying that 90% of all the barns have been enlarged and modified at some time.

Even the barns of the Lancaster, Pennsylvania area that date from the time of the Revolutionary War are being updated, changed or added onto as the current need arises. Driving through the area, in addition to ones

The American Barn is alive and well and living in Amish America!

Detail of interior of barn described on page 22

Below: An example of a home from the Rhine Valley in Germany, using oak beams in the walls. This technique was adapted with changes to the construction of the American barn.

easily recognized as early stone barns from around 1800, I saw many that had been modified, with tin siding covering the stone structure and all that I saw during one day had had the forebay or overhang extended for additional stable space, for both cows and horses.

Heyl suggests that the cantilever or overhanging forebay type barns bear a distinct relationship to medieval structures in the uplands of Rhenish Palatinate and the Alpine heights. It should be pointed out that these are the origins of the Amish and the language they speak today is close enough to the broad accent of the Palatine region to be easily understood. I learned this from a German radio broadcaster who taped interviews with the Kalona, Iowa, Amish, saying their dialect would be easily understood in his region.

So the Amish brought not only their language, but their knowledge of building with oak beams to this country since barns as they exist in this country had no counterpart in the old country. It was natural that the structures should be patterned after castles, city walls, military ramparts, and even the way homes were built on steep hillsides. The slits or holes in the barn walls suggest the holes used for shooting arrows and later small arms from town walls or castles. In the barn these slits served the purpose of ventilation and were later replaced by small doors (such as a hay mow door), which could be used to load hay, then be closed off during the worst weather. Ventilation of hay storage areas was important, since gases were formed that could lead to spontaneous combustion or ignition by lightning. The method of framing wood timbers interlocked with wooden pins can be seen in many German town today by simply going into the back alleys where the structure may still be seen undisguised by a new false front.

It was to the Amish and other German people with the willingness to sacrifice their country, their homes, and their lifelong friends for a chance at a new life in a new world that led to the creation of the American barn and the American farmstead.

The opening of Pennsylvania to settlers under the watchful eye of William Penn and other Quakers allowed oppressed people like the Amish to escape crippling social and economic pressures. The Amish were burdened with taxes, tithes, and service to the military of whoever was then in power, or yet more taxes for not serving. Pennsylvania beckoned with fertile land and religious freedom. Fertile land seemed without limit.

To produce in abundance depended only on a man's ability to work the land. To the land hungry peasant of Europe, it was an answer to prayers, even though many had to literally mortgage their lives and their labor as indentured servants for as many as six years. These people actually put themselves into the hands of a ship's captain to be sold upon arrival.

From an arrival in poverty, these Pennsylvania Germans, Amish, and others, labored hard to build a place for themselves. So well did they succeed that in a period of a little more than 50 years, the Lancaster area had become the breadbasket of America, supplying most of the wheat and other essentials that had often heretofore been imported.

A report from 1787 described the average farmer as compared to the Germans: "The average farmer possessed one miserable team, a paltry plough, three acres of Indian corn, as many acres of half-starved English grain from half-cultivated soil, with a small spot of potatoes and a small yard of tillage...Pastures are never manured and mowing land seldom...This compared to the well tended farms of the Pennsylvania Dutch which were not for sale or were too expensive for the average immigrant."

In other words, the Pennsylvania Dutch (Amish and other German farmers) had become the farming elite. It was the Amish who manured their lands, composted and diverted streams to irrigate and used various other methods to upgrade their land, not the least of these which was building a series of magnificent barns. I would estimate that no less than 200 barns built 200 or more years ago in Lancaster and nearby counties still stand.

Near Myerdale, Somerset County, Pa. Barn raising from second major settlement of the 18th century. Today this area has few Amish although it supplied many who migrated westward to Ohio, and then on to Iowa. Photo supplied by Glenn U. Suder.

Dover/New Philadelphia Times-Reporter

Raising a barn

By BRIAN WILLIAMS

Though heavy snow had piled up and weakened the structure, most of it had been washed away by the time the blizzard of Jan. 26 caused the roof of a unique, historic eight-corner barn to collapse.

Thursday, about 200 Amish friends and neighbors of Mrs. Verna Miller gathered at the homestead on Holmes County Rd. 172 and began raising a new barn in the frigid 10-below-zero morning air three miles northwest of Walnut Creek.

Shortly after noon, just four hours from the time they started, the men had already been fed their first meal of the day by neighboring womenfolk and were back to work, straddling high beams of the skeletal structure, hammering siding to the frame, carrying, marking and measuring lumber and tossing it up to their co-workers.

But two anachronisms stood out in the systematic scurrying: a hydraulic front end loader and an electric saw were used to help erect the new, four-corner barn where the old, many-sided one, the only one of its kind in the area, had regally rested

"IT WAS as old landmark," said Mrs. Miller. "It just doesn't look the same anymore — the new barn seems so much closer."

She said she thinks the old barn was built in the early 1900s, but, "I don't know that there's anybody living yet that remembers when it was built."

"We thought about building the new one like the old one, but people talked us out of it," Mrs. Miller said. She said it would have been very costly to put a roof on an eight-sided building.

One workman said the sides of the

all in day's work

old barn were in fairly good shape but that it would not have been practical to rebuild it like it was.

Mrs. Miller said her oldest son, Willis, hasn't been to work at Schrock Woodcrafts in Walnut Creek since the roof collapsed. Instead, he's been in charge of cutting down trees and sawing lumber for the barn-raising.

"We have a sawmill in our woods and we cut all the trees ourselves," Mrs. Miller said.

Willis said he's had about 25 or 30 volunteer helpers every day. "We started sawing two weeks ago Monday," he said. In all, about 125 trees had to be cut down to get 25,000 board feet of lumber.

THE BARN-RAISING was not a totally self-sufficient operation, however. The Millers bought siding for the new barn, with the siding from the old barn used as flooring for the new structure. Willis added that some neighbors tore down an old wooden silo and that wood also was used as flooring.

Except for some support posts in the basement, no other wood from the original building was used.

A man overseeing the proceedings, who downplayed his own involvement, ("It's the Lord's hands that's doing this."), said he hoped the roof would be on by 4 in the afternoon.

By that time, Mrs. Miller and her family said the men would be hungry again. The women would quit quilting to heat up food brought in by area churches and prepared by many neighbors.

Then it would be back to work, in hope that the job would be done by the day's end.

"We owe the community a big thanks," Willis Miller said. "We couldn't have done it ourselves."

Barn raising near Berlin, Ohio. Photo by Oscar Miller.

The Amish put no lightning rods on their barns, feeling to do so would be to show a lack of faith in God. But it is not heaven who helps the Amishmen re-build the barn, but his fellow Amishmen.

Not only a lightning strike or tornando can bring down a barn, but carelessly put up hay which in spoil turns to methane It can spontanteously combust during the hottest August weather.

THE AMISH WAY

David's family were all
awake, and dressed, waiting out
the storm, half asleep
when lightning hit the barn
and tore into the hay-filled mow.
The fierce, fiery furnace of flames
beaconed neighbors to the scene.
Ladders and bucket lines did save
their house, the rest burned to the ground.
More cows, horses, chickens and pigs
were lost than could be led away.
David, Sarah, and grandmother Blank
comforted the children, and said
a prayer of thanks that all were safe.

Boys on horseback spread the news
that brought the friendly help
of men who cleaned all trash away
around the standing fieldstone walls,
and women set some tables full of food.
The preachers (farmers, chosen by lot)
announced the time when all would come
again to raise the barn and sheds.

On that day the teams arrived
at dawn and soon the men, like ants
upon the rising structure, turned
the burned-out place into a barn —
with pens, and coops, and carriage house.
The only ones who did not come
had been assigned to daily chores,
and soon they brought their tools
to join the busy working force.
All day good food was served.

Pigeons are back to coo near the door,
Children are running across the new floor.

<u>Sweet & Sour</u>
Poetic insights into the Amish way of living.

This little book has never gotten the attention it deserves. The poems are by Alfred L. Creager, 32 years a minister of Trinity United Church of Christ, Collegevill, Pa, a college professor and friend of Amish. They mesh perfectly with the pictures of William K. Munrio, of Riverton, N. J., illustrator and book designer with an eye for the essence of Amish. Delicate, insightful, and humorous describes both words and drawings. Copies of the $6.00 book are available from Amish Heritage, listed in front or from: A. L. Creager, 139 Seventh Ave, Collegevilla, Pa 19426. Phone (215) 489-2677.

Barnbuilding near Nappanee, Indiana. Photo by Freeman Borkholder.
Opposite page: Barn raising near Wakarusa, Indiana. Photo by Wakaruse Tribune - Judy Martz and Shirley Pitney

Barn Raising

The American barn weathered many changes its time. In the beginning it started out to be a simple rectangular box, simple enough to be built by simple men with simple tools. Despite all the fancy designs and styles, the window and door designs, the cupola, and the different roof lines, the barn had only one basic purpose-to preserve the grain and the livestock so that the farm could prosper.

In today's world of pre-fab and pre-cut, it is hard to realize what building a barn entailed. It started perhaps with a survey of nearby woods (even today, the Amish prefer land that has 5 to 10 acres of woods). Trees to be used for the building were notched to identify them and would later be felled with a good sharp axe. Then it would dragged by horse or oxen to a nearby site where the timbers would be cut and stacked in preparation for the building of the barn. Many barns contain hand hewn beams, done with nothing more complex than a broad ax and an adze. Some of the oldest boards in barns were

Near Kalona, Iowa. These Amish use a new style of barn building - concrete block with a gambreled tin roof.

hand sawn with a gang saw, a series of up to five saw blades that could cut as many as four boards at a time for siding. Examine any barn today built in the early 1900's and chances are, inside you will discover hand hewn beams and boards from yet earlier barns. Six by eight inch oak beams are often recovered and re-used. The Amish today have been known to dismantle older barns in order to re-use its parts for ones they are repairing or building elsewhere. Anyone examining the lumber would see that it is much better than most of the lumber obtained today.

Many old barns still stand today, because they got the best possible site for construction. The barn needed high ground so that wastes could drain off. It needed a source of fresh water nearby. In the early days, there were no windmills to aid farmers in bringing up water from underground. The early ingenious Amish in Lancaster rigged water wheels in nearby streams to pump water and run machines and these are still in use throughout Lancaster.

Raising the huge timbered barns required more manpower than an ordinary farm had available. So when

Above: Modest frame barn raising near Jamesport, Mo, first of over 12 Amish areas in Missouri, begun in 1953.

the time came you called on neighbors for miles around. The women cooked plenty of food and they would make a kind of holiday out of it. The day's end would see a barn in place and everyone would have the feeling that when his turn came, his neighbors would step in and help him. The barn raising engendered a sense of mutual trust and responsibility among neighbors. Although they often lived miles apart there was more of a sense of neighborliness than one can find today in an apartment complex in a big city. The barn raising and other community activities fostered a sense of brotherhood that is totally lacking in society today.

Only the Amish today retain the sense of brotherhood that makes them travel across the country in mass to help victims of tornadoes and floods.

The tools of Josie Miller, barn builder. At left a small adze and the chisels used to work on some of the peg holes. Next is the hand crank that was used to drill the holes for the wooden pegs, and next to it, a new gasoline powered drill to get those pesky peg holes drilled. It was from this sort of joining that the expression "square peg in a round hole came about."

Josie Miller, Barn Builder

A short stout tree stump of a man, gnarled of face and hands, Joe T. Miller, often called Josie by his friends, is an Amish barn builder extra-ordinary. In his nearly seventy years, Josie, who lives just outside, Kidron , Ohio, has built, by his own reckoning, between 300-400 barns.

"Over 400!"' says his wife. Josie doesn't keep count. "Josie as a young boy was not much taken to farming," said his wife. "He liked to fool around with tools and make things of wood." He was taken under the wing of her uncle Dan Weaver, who built barns throughout his life.

Josie described him as a patient man, very painstaking in how he described how to do a job, but when he walked away, he would turn back and say, "Now that you've been showed, you'd best not ask again."

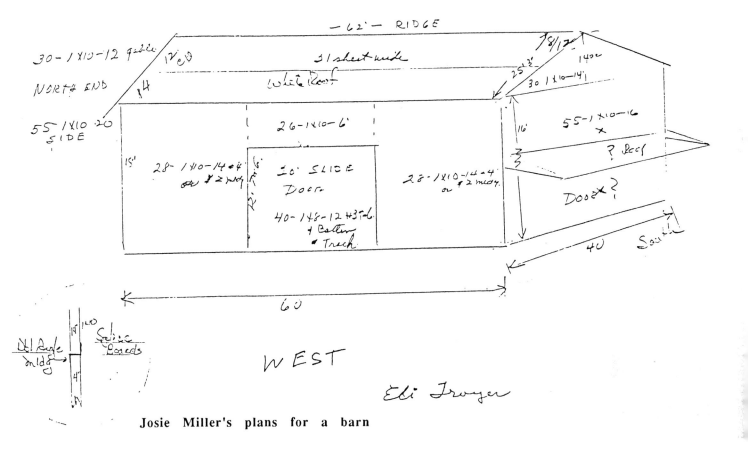

Josie Miller's plans for a barn

To listen to Josie Miller talk about barn building is a lot like listening to a sculptor who must direct a team of workmen in casting a monumental piece, or an orchestra leader preparing to conduct a major piece. Preparation for Josie consists of visiting the site where the barn is to be built. Then he makes a preliminary sketch on the back of an old calender sheet, the kind of calenders that are given out by the lumber yards each year. He has to take into consideration the lay of the land, or in the case of a barn burned down, what the original dimensions were and if they should be changed.

"Nearly every builder builds a little different," Josie said. "I like to get to get the job in my mind. I don't talk about it, I try to visualize what the finished barn will look like." Josie's barns are built like the Amish barns of more than 200 years ago, of strong notched oak beams joined with wooden pins which will not rust away, as surely as any sculptor Josie is building his barns for both today and tomorrow, two hundred years from now they may still be standing as his monuments to good farms.

Like the true artist, Josie suffers from butterflies in his stomach before a performance. The night before a barn building, he cannot sleep.

"You try talking to him the evening before and all he does is grunt," says his wife.

"It's a worry," says Josie. "When you're out at a barn, you can't think of something else. You can't doubt when you lay out a barn, you have to be positive. You've got to set your heart to it."

An Amish barn building is no slow affair. It begins at seven in the morning with a pile of lumber and is usually finished by the end of a day, with some barns over 100 feet long.

Others have said that Josie may seem small. He has a big voice on the job, and again like an orchestra conductor, he has to direct his trumpeters to raise the walls, then his drummers to finish the floors. The speed with which Amish barns are built depend on teams of experienced builders who have specialized in the framework, the sidewalls, and the floors. Josie Miller wields the baton of his voice until the last board has been nailed in place.

On these pages are the plans of Josie Miller. Some of his sketches of barns have been framed and hung in an architects office, but no architect is he. A simple wooden tool box contains most of what is necessary to drill and notch the large oak beams that lock the structure together as sure as if it had been built with steel beams and rivets.

Sighing, Josie Miller comments, "Old farmers were just as clean with their barns as women were with their houses," as if he regrets that once his sculpture is done it will be necessary to soil it with use. The least they could do, he seems to say, is keep the floor swept as they used to.

The whole Miller family is interested in barns. They told me of what they thought was probably the oldest log barn in Ohio, but unfortunately between the telling and my getting to photograph it, it was bought by a man from New York State, dismantled, and prepared for shipment there.

Although Josie's wife expressed the hope that Josie will slow down, she doesn't hold out much hope, "They are after him all the time," she says. "He will be building barns till he dies."

Photo by Oscar Miller.

Above: Octagonal barns are rare. When this one burnt from a farm near Walnut Creek, Ohio, Josie Miller was called in to replace it twice - the second barn got struck by lightning. Below is Josie's replacement.

Barn above near Kidron, Ohio. Barn below west of Wooster, Ohio just off Route 3. It was built for the Swartzendruber Amish, a very strict old order group. They bought a large farm and divided it into five and had Josie Miller build five barns.

On the streets of Shipshewana, Indiana a hybrid barn built of huge oak beams and fitted with pegs. It is a gift shop in a town that has exploded as an Amish tourist area.

Big Dave Bontrager, Barn Builder/Construction

Big Dave Bontrager well deserves the name Big Dave. His hands are as big, it seems, as baseball gloves and each finger looks the size of a Polish sausage. Those are the hands of a barn builder, one who can wrap his hands around a six by eight inch barn beam and make it seem no more than a two by four in the hands of an ordinary man.

The day I met Big Dave, he was wearing, as they say, two hats; one as barn builder and the other as head of Bontrager Construction Company. That day he was building a barnlike structure in downtown Shipshewanna, Indiana. It is to be a new gift shop built as if it were an Amish peg barn. Bontrager recently completed a similar structure for a new restaurant in southern Indiana, in Davis County near the town of Montgomery. He also did a barn near Dayton, Ohio for a state park. More and more his skills as an Amish barn builder are being called into play for general construction. He has been in the barn building business for 15 years, and has built between 20 and 25 barns at Amish frolics (the dictionary defines the word: to play and run about happily). The Amish frolic gives the Amishmen a chance to visit and to test their skills against one another. A barn raising is a kind of work/play situation. They might compete to see who is the fastest at nailing on the roof or at getting the sidewalls into place. It is a kind of fun for a good cause, because usually the farmer has lost his barn to fire, tornado or some other disaster. Big Dave is following in the footsteps of his father, Joseph A. Bontrager and Uncle Dave A. Bontrager, who built barns together for more than 32 years. They were responsible, he believes, for well over 100 barns.

In this building, Big Dave is using his construction crew who have built all kinds of structures, but they are constructing it as if it were a three story tall oak beamed peg barn.

All the oak beams were cut at the sawmill on his farm, from locally cut oak timbers. Bontrager is one of a new breed of barn builder carrying on the tradition of his father, but making a living in the construction business. One notes that the Amishman, to survive and prosper is anything but lazy; he farms, he cuts timbers for barns at his own sawmill, and he is in the construction business.

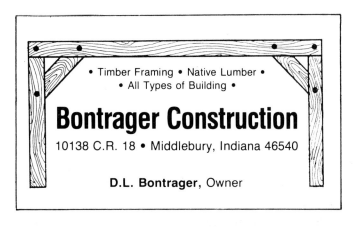

Kalona, Iowa, Sunday church service.
Opposite: Near Nappanee, Indiana, these Amish use the barn for their church services as they have since the beginning of barn building in America. All farmers once held church services in barns.

The Barn as a Church

Like the early pioneers, the Amish hold church in the barn. The former because they hadn't yet built churches, the latter because they build no churches. In the winter, the home is their church, in summer they use the cleaned out threshing floor of the barn.

John A. Hostetler, in an early thesis paper I uncovered in the Goshen College Library Historical Section describes a baptismal service in a large barn near Middlebury, Indiana on what he describes as a beautiful September morning. "The sun shown on the faces of the audience through the entrance of the large swinging doors, propped open on the back of the barn. The clear blue sky and the warm sunlight were a symbol of the special occasion of the morning, a baptismal service."

He tells us that as the songleader begins to sing the first song, the ministers, bishops, and deacons retire to the house to discuss ministerial matters and the coming baptism.

Meanwhile, the farmer in whose barn this is taking place runs around opening a second barn door behind the women's seats, and finally another one in the straw shed as the day heats up. Below, the horses stomp in the stalls while outside the yard is packed with black buggies.

The congregation is segregated according to sex and age, seated on flat, hard benches with the women on one side and the men on the other, older toward the rear, younger toward the front, married and unmarried.

The ministers, deacons, and bishops enter in leisurely fashion and proceed to shake hands with all who are nearby as they make their way toward the minister's bench.

After singing and a number of introductions, the bishop takes over for an hour long sermon before addressing the applicants for baptism. The deacon leaves and returns with a small pail of water and a tin cup. An older woman assists with the removal of the headcovering from the women to be baptized and the

Funeral at Walnut Creek.   Photo by Alma Hershberger

deacon pours "three drips of water" on each head. The head coverings are then replaced.

Hostetler concludes, "The water dripped from the head and nose. But it had to drip because the applicants were in deep meditation...The pure white aprons were saturated with water, and the fringes were soiled with the moisture and dirt of the barn floor. Overhead high on the inside of the barn, the pigeons were flapping their wings as they flew from one end of the barn to the other. A gentle breeze brought from the open door of the straw shed a cloud of find particles of chagg and dust. High in the clear blue heavens an airplane roars in the distance, a symbol of earthly wisdom, progress, and evil."

Note that Hostetler has taken on the Amish persona, here to equate, in Amish fashion, progress with evil. The Amish hold fast to their isolation and the barn as church is one of their most cherished customs. It brings to mind baby Jesus sharing a stable with animals of the farm and field.

Near Strasburg, Pa. These Amish girls enjoy rowboating during a typical Pennsylvania style Amish hoedown. See page 51 Amish girls cross dress as guys during a dance.

Barn Dances

The American barn dance is as much a part of American farms as all the many labors that farming entails. To get an account of a barn dance, one usually has to go to a grandmother, but today it would be possible to talk to Amish teenagers in any one of 22 states and Canada to get a first hand account of the goings-on at the last barn dance. The lady who has written this piece was not too many years in the past an Amish teenager attending barn dances in Holmes County, Ohio. Alma Hershberger is an ex-Amish lady who is willing to tell of her younger days in hopes that people will better understand the Amish. She has done a book on her own childhood, two cookbooks and is collaborating with me on "Amish Women Across America". She will supply the insights and quotes from anonymous Amish friends, and I will supply the pictures. She has also agreed to put back on an Amish outfit (for educational purposes only), to speak of the Amish way of life in schools and on television, if there is an interest. Her books are available thru Amish Heritage Publications and are listed in the front of this book.

Barn between Walnut Creek and Winesburg on Route 515. Location of a barn dance. Photo by Alma Hershberger

Barn Dances by Alma Hershberger

You "English" who think we Amish are pretty straight-laced folks have not seen us break loose at a barn dance. Old folks talk about young folks sowing their wild oats; a barn dance is a fine place to sow a few. If you'd come to an Amish barn dance you'd really see we can shake a leg with the best "English" teenagers.

No rock and roll for us. The guitar, the fiddle, and the harmonica provide the music. A good deal of hand clapping and foot stomping help to fill out the sound. We tend to like countrywestern sounding songs. The old ones like "Turkey in the Straw" and other square dance tunes are pretty square now a days. Most of that has been replaced from songs off a countrywestern station. How do we hear about them you asked? Even though we don't have electricity, it doesn't mean we can't hide a portable radio in the barn for a little music to milk the cows by.

Barn dances are not a regular thing, but tend to be on the spur of the moment, when the parents have gone off for a day or two to visit relatives, preferably in Canada.

People wonder how we hear about these barn dances, but the Amish telegraph is really quite good, especially if you keep your ear to the ground. Folks say that if a pin drops on one end of the county in the morning, you can hear it echo at the other end before nightfall. Most of you know we don't have phones in our homes, except the New Order Amish, but that doesn't mean we don't know how to use them. Just down the lane from the farm or just down the road apiece there's a pay telephone. We wouldn't want to be without a phone to call a doctor in case a baby was coming early or there was some kind of emergency.

My sister and I would get the word: "Barn dance at so and so's place," and after dinner and chores were done we'd get ourselves together and hike up the road where the guys would sometimes pick us up in their cars. You didn't think we had cars? Well, the guys haven't officially joined the church, so they can get by with having one tucked away. You should see the cowboy shirts they like to wear. We don't always go just plain either. Sis keeps a little make-up and lipstick. A little spot of color would do wonders for a girl's ego.

Now all of us don't come in cars. There are always plenty of buggies and by the time dark is coming on, the barn is swept out, and the buggies and wagons and other farm equipment moved off the threshing floor, usually hauled out to the barnyard. A kerosene lantern or two are hung from the rafters.

We do have beers and softdrinks and occasionally one of the boys can get kind of foolish. Sometimes if the horse didn't know the way home, the boys would never make it. One of the guys laid his guitar down careless like as he was leaving and his horse stepped on it. Sometimes we don't get home until time to begin the morning chores, but by then we've blown off so much steam we are set to roar for the rest of the day. It's a good feeling to do foolish things once in a while; let your hair hang down as you "English" say.

Usually it doesn't make any difference how much noise we make since the old folks are almost always gone, and we're far enough away from the neighbor's farm that they can't hear us, even if we make enough noise to raise the dead. One time though, I went to a barn dance at a place where the parents hadn't gone out of town.

Most of us were gathered at the back of the house with a walk out basement and a large overhanging open porch above. A lot of boys and girls showed up, more than anyone ever expected. Things can get pretty noisy with all the laughing and talking. Of course, there is always one in the crowd who's voice is louder and carries over everybody else's.

The father was upstairs in the house and he couldn't sleep for the noise. This young boy's voice finally got to him. He got up out of bed to see where this voice was coming from. Then he went down to the kitchen cupboard and filled a container with water. He tiptoed to the edge of the open porch and poured the water down over the ornery boy. This quit the loud mouth with a lot of laughter following. Everybody got the message to quiet down. Most young folks didn't know that the old folks were at home.

NOTE: This kind of high spirited and physical play, along with practical jokes and horse play, are as much a part of this pioneer ritual as is the music and dancing.

Barn games

John Hostetler's book Amish Society, contains some of the most detailed descriptions of the barn as a social structure in Amish life. It is not only the Amish church, the town meeting hall, the funeral parlor, but also a workshop and gymnasium as well.

Hostetler describes barn games which he says are commonly played at wedding and husking bees. He talks of weekly hoedowns getting out of hand in larger Amish settlements. He names such games as "Skip to Ma (My) Lou and "Six Handed Reel" and describes them as party games that involve holding hands and swinging partners.

Near Topeka, Indiana. Children get involved in haying.

Amish Children and the Barnyard

The barn and barnyard for Amish children are a microcosm of the world. Here they can watch the mother hen and her chicks, the mother cat and her kittens, the mare and her colt, the cow and her calf. They learn that all creatures large and small are nurtured by their families, and that all things large and small have their place in the world.

Playtime gives youngsters the chance to explore the rafters, haymows, stables, pens and granaries; they can climb to dizzy heights in the rafts and route the pigeons and barn swallows, or crawl thru the hay to locate the hiding place for mother cats new litter of kittens. The barn is full of sights sounds and smells that a child never forgets.

The land around the barn is almost always a ferment of activity from dawn to dusk, and often into the night hours by lantern light or early on a winter morning, before first light.

The barn and barnyard is always filled with pieces of horse drawn equipment, from plow to manure spreader. It is always time to do something around and Amish farm from earliest spring right on into the deepest winter.

On a prosperous farm (one with an abundance of little helpers), the place can be alive with activity. Each child is started at an early age to learn his or her chores. The youngest might be given the job of feeding the chickens, ducks or geese, or gathering their eggs. After a time a young boy might tend the horses after a hard days work in the fields. They need to be watered, feed and rubbed down. The girls graduate to hoeing the garden and later helping with washing, ironing, and the canning.

I knew an Amish girl who felt literally chained to the barn. From the time she was eleven until she was 21 she had to milk from six to eight cows morning and night. At 21 she found a nice non-Amish man who had no cows and married him.

Amish life is altogether from another time. Lantern light provides the only illumination for late night and early winter morning chores as well as for reading, sewing and other household chores. From the earliest crowing of the rooster, the barnyard is a beehive of activity. Horses must be hitched in preparation for the day's work, the cows must be milked and put out to pasture for the day, and the hogs must be fed. A windmill pumps away with a slight creaking noise, filling the horse trough with surges of splashing water. It is time for another day's work, and if still during school time, the children must get their work done before going to school.

The Barn as a Factory

The farm of the last century was no single crop enterprise, but a diverse system whose bottom line was to sustain the family, despite reverses in weather or in the economy.

Chickens and poultry ran around the barnyard of yesterday's farm surviving as best they could, while waiting for the day when the farmer would receive a postcard notifying him that poultry of all kinds were being bought at the local railroad siding. Then it was "goodbye chickens!". An Amish farm keeps laying hens only through their peak production time. Then he cans them or makes chicken soup out of them.

Like the farmer of the past, the Amish can every food possible. They keep a collection of canning jars that go back to great grandmother's time, canning everything from apples to zucchini. They grow their own wheat, oats, and barley and are capable of grinding it themselves if necessary. They keep bees for honey, tap maple trees for maple sugar, grow sorghum to produce molasses, and usually have a sizeable orchard to back up their vegetable garden. One Amish woman in the Kalona, Iowa area with the aid, one assumes, of many daughters, canned over 2,000 jars of fruits, vegetables, and meats in a single season.

No single crop dominates the Amish farm. Corn, wheat, oats, barley, and tobacco can all be grown as cash crops when necessary.

In addition to the barn proper, there are often many other useful buildings. There is the wood shed (of which Amish boys are all to familiar in more ways than one), tool shed, chicken house, rabbit hutch, smoke house, root cellar, ice house, and sometimes woodworking or repair shops. A number of Amish skilled in building their own homes and furniture have turned to sidelines of construction and cabinet making.

For those farm families of yesterday, it was a matter of survival against the elements of nature and the economy. For the Amish, a third element has been added-the battle against the encroachment of the outside world. The work has to maintain their plainness in the face of a constant stream of new and modern temptations.

One of those temptations is the tractor, there are now old order Amish who are using tractors in limited fashion. They have never maintained that some mechanical devices don't have their uses and gasoline powered hay balers can now be found in many Amish areas, pulled of course, by a team of horses..

Today barns on non-Amish farms have become little more than storage bins, and resting places for tractors and combines, with an

Near Kalona. Amish milking machine

old sleigh or wagon tucked away in some corner gathering dust. On the Amish farm this equipment is still very much in use. The Amish of Hutchinson and Yoder, Kansas are wheat farmers forced to switch from horses to tractors and combines because of 100 degree heat under which wheat is often harvested. The minute the wheat has been harvested, it is loaded into horsedrawn wagons for the trip to the grain mill.

As the Amish continue to spread across the farming country they select the most remote regions of farm states and the old barns still standing are being dismantled and re-assembled on a new site.

The Amish continue to demonstrate that manpower and a willingness to work as a brotherhood can bring them prosperity and bring old farms and old barns to new life.

Route 133 at Interstate 57, east of Arcola, Illinois. Note the Amish horsedrawn hayrake in front of newly painted barn.

Mail Pouch Signs

For years as a photographer, I have been intrigued by, and often stopped to photograph those barns covered with the Mail Pouch Tobacco advertisements. For this book I decided to track down the origins and find out who was painting them. The story begins with the Bloch Tobacco Company, who began painting these signs on barns around the turn of the century to advertise their products. Most of the barns were in the Midwest.

Over 20,000 barns once bore this sign, as far away as California and Oregon. Today there is but one painter of Mail Pouch barns and that is Harley E. Warrick. He joined a painting crew in 1947 and since 1965, has been the only man left on the job.

Harley almost lost his job in 1966, thanks to Lady Bird Johnson's Highway Beautification Act. For a while it looked like he would have to hang up his brush but a furor arose and a sympathetic U. S. Congress declared them National Landmarks.

In Winesburg, Ohio. One of many Mail Pouch signs to be found in this Amish area.

Harley estimates that he has painted 10,000 barns over and over again. Today about 1,000 barns in Michigan, Illinois, Indiana, Kentucky, West Virginia, Ohio, Pennsylvania, New York and Maryland still bear the fading signs. The General Cigar and Tobacco Co., of Wheeling, West Virginia now owns the business. A spokesman for the company says when Harley lays down his brush for the final time, he will not be replaced. Today the company has only two hundred leases on barns in the Midwest.

Harley says he would keep right on painting those signs, even if the tobacco company decided not to pay him anymore. Not painting them, he says, would be like letting the Statue of Liberty fall down.

If Harley Warrick never painted another side of a barn again, he would still keep painting the Mail Pouch sign for as long as he lived. Harley has enough orders from people who want him to paint small replicas on barn board, that he swears if he lived to be three hundred years old he couldn't paint them all. His greatest honor was to be invited by the Smithsonian to paint Mail Pouch on a wall for the 1967 World's Fair in Montreal.

Route 62, south of Millersburg, Ohio.

"I get requests from all over the country," he said. "Up there in Akron (Ohio), I've done about 12 family rooms. I've done some of them cowboy bars, too."

"It don't make no sense to me," he chuckles, "when something gets scarce, Americans want it more."

Harley Warrick has painted so many Mail Pouch signs he swears he could do it in his sleep. The strangest request of all came from a circus that asked him to paint the sign on the side of an elephant. "It was a little like painting on burlap," he laughed. He did the job in water colors. No problem with the elephant, he says. "It was very obliging."

So if you should chance down a road in an Amish area of Indiana, Illinois, or Ohio and see someone painting away at a Mail Pouch sign, give him a wave. Chances are it is Harley E. Warrick, preserving another National Landmark.

On Route 39, between Walnut Creek and Sugar Creek. This is the oldest form of Amish haying machine; a kind used by all farmers up to the 1930's.

The Horse and the Barn

The American barn is what it is today because of the horse. Without the horse and to some extent cattle (oxen), the barn as it stands might never have been built. The English farmer kept his horses in a lean-to and stacked his hay in haystacks.

The German farmer coming from the Rhine Valley and the Alps knew harsh winters when cattle and horses might be frozen where they stood if not given proper house.

In Lancaster, these Germans discovered a weather similar if not quite as harsh, as in the high Alps. The answer was to build suitable shelters to house, what in America became growing herds of livestock and horses, since there were no kings or dukes here to requisition the horses for a new war, or for the tax collectors' transportation.

Near Alymer, Ontario, Canada. Manuring and composting is the Amish way of enriching the soil.

Above: Near Alymer, Ontario. A young Amish boy takes his turn discing the field before school.
Below: Near Belleville, Pa. This Amishman is plowing in Big Valley, the third Amish settlement in America. This valley is home to the white buggy Amish, who live and dress in the styles of colonial times. Women wear no bonnets, men no suspenders.

Above: Near Kalona, Iowa. This horse is one of many in the area being bred by the Amish for sale, rather than for their own use. In recent years, there has been an increasing demand and draught horses.
Below: Near Topeka, Indiana. An Amish boy, 11, cultivates the corn, while his younger brother watches and learns.

The Barn as an Icon

Definition of Icon from Webster's New Collegiate Dictionary: Object of uncritical devotion.

Lewis Evans, 1753: "It is pretty to behold our backsettlements where the barns are as large as palaces, while the owners live in log huts; a sign of thrifty farming."

Scratch even the most dyed in the wool city dwellers in their high rise apartments and chances are, underneath somewhere you will discover farm memories. They may never have left the city, they may find it hard to recognize the difference between a horse and a cow, yet somewhere in the memories lurks a farm heritage. In Jefferson's day it took 85 percent of the population of this country to feed the nation. Today it takes less than five percent of the population to feed not only this nation but to ship grains around the world.

So those who had farmed the land had to find new occupations in cities and in factories. There is an old saying: "You can take the boy out of the country, but you can't take the country out of the boy."

The country may be three generations back but chances are grandma or grandpa talked about the good old days on the farm. Somewhere hidden behind the long years of city life and the stink of garbage and the smell of carbon monoxide is the faint memory of the smell of new mown hay, horses, and manure and a faint image of a red barn.

I am definitely a city boy born and bred, yet I have had my time in the country, climbed up into the hayloft and jumped down, stepped in a cow pie, and raced along between rows of tall corn in the hot August heat. I have been drawn back to rural life again and again after living around New York, Chicago, and Kansas City.

If we had no first hand acquaintanceship with farms, we have seen it portrayed in all its wholesome glory in countless motion pictures and television programs. If we are observant enough when we walk the residential neighborhoods of urban America, we will find ourselves surrounded with copies of the American barn. One called the German Barn or gambreled roofed barn was transferred almost wholly to the American house, the saltbox style of barn became the saltbox house of New England. Today the copies are much more marked an imitation, with board and batten siding made to resemble barn siding, and now there are some house set up to look just like barns.

It is increasing in vogue to tear down a barn and build it back up inside out in the living room or den of a house, or in a local business or bar. Who has not visited someone who has done their walls or kitchen cabinets in weathered grey barn boards? Ads in the local shopper's want ads offer for sale barn boards at one dollar per foot, either faded red, or greyed and without paint. And if one did want barn board, how about a couple of 40 foot long hand hewn oak beams?

Barns are as much a part of the American heritage and the American dreams as castles are in Spain. Childhood memories of the excitement and wonder that the barn and barnyard represent pervade the American psyche. You can buy a little pre-fabbed hipped roof barn in wood or metal at your local hardware store; perfect for storing your lawnmower and gardening tools. You can buy barn shaped doghouses, mail boxes, and bird houses. You can buy wall paper with barns and farm images for the kitchen along with a barn shaped bread box. In fact, no matter where you live chances are you will see something relating to American barns almost every day.

The photograph on the opposite page is my contribution, along with others in this book to barn iconography. It was done by special insistence of my publisher who wanted an ideal technicolor dream of a farm for the cover of my book Portrait of Iowa. With nothing but a cover and a title, my publisher ballyhooed the book across the state before I even knew fully what was going between the covers. It sold over 6,000 copies in the first six months and today, 15 years later, still continues to sell with the same cover and revised insides.

Numerous requests were made for this photograph and the University of Iowa Hospital had the largest color print ever made from this photo to hang in the Board of Regents meeting room. Apparently it struck many as the quintessential view of Iowa. It had everything that said Iowa, green grass, red German style barn, working windmill, blue sky, and horses. I had driven hundreds of miles in search of just such a scene and found this one on an Amish farm only a few miles from my home. The Amish children just out of camera range assisted me by driving the horses into the forefront of the picture.

To me it suggests the peace and tranquility the American farm once had. For it was not the adventurers and explorers who tamed this vast American landscape, but the American farmer who put up his barns from sea to shining sea.

Near Arthur, Illinois. This sunlit farm scene is backed by dark ominous clouds of an impending thunderstorm.

The Weather

    A man stands alone with the horses in the midst of a large hayfield on a hot August day. He awaits the coming of the next wagon from the barn. It has been a hot sweaty afternoon and the hay was almost gathered.

    He could have been the last man on earth, so isolated was he in the pocket of the hills . Off in the distance huge thunderheads were forming. They might dance in the sky all afternoon and do nothing, or suddenly like a roaring freight train come racing across the landscape bringing a solid sheet of rain. That would mean the hay would have to be turned again before it could be gathered and stored in the barn.

Near Charm, Ohio. A typical farmhouse and barn layout, with both showing many additions.
Next page: Huge Ohio barn easily seen from the parking lot of Walnut Creek Cheese House, Route 39, Walnut Creek.

Such clouds, if joined by cold air might quickly become a cyclonic wind or a tornado that would tear across the land removing farm and barn and all in its path. Even a lighting strike from a swift moving thunderhead can kill a man in the field, as it did recently Terry Dixon, a leading Iowa farm activist.

Wind and rain, sleet and snow, every twist of the weather can mean another danger. But the Amish and their sturdily built barns have learned to live by the season and not by the clock. Many of their barns have escaped both the havoc of weather and the neglect of man to stand as firmly today as when they were built 200 and more years ago.

On Route 45, East of State College, PA.

Round Barns

This round barn was built for Calvin Neff. Neff had visited the middle west some years earlier and was impressed by the round barns he saw there and believed the design was adaptable to Pennsylvania farms. He himself drew up the plans which called for a A round banked barn, with a huge beam substituting for a center silo. The center pole was 56 feet long and hewn from a single tree to a square 14 inches. Using the pole as a focus, the outside foundation was built from farm-quarried limestone, with the north section of the wall sufficiently high to provide entrance to the second story. Lumber came from a stand of virgin pine nearby and 100,000 board feet were sawn by a local sawmill for $300, with the timber costing $100   The sawn lumber was moved to the Neff farm the following winter by horse-drawn sleds. The barn shingles were sawn on the farm from heavy white pine sections cut to shingle length. Over the years, many changes have been made but the barn still stands and has become a local tourist landmark.

Near Strasburg, Pa. Amish girls have fun on neighbor's pond at a hoedown on Saturday evening.

Round barns have a unique character all their own. One of the oldest surviving round barns is the one built by the Shakers in Hancock, Massachusetts in 1824. It was said, perhaps tongue in-cheek, that they built a round barn so the devil wouldn't have a corner to hide in. Round and octagonal or polygonal houses were built in the late 18th century. President Thomas Jefferson built an octagonal summer house in Virginia in 1806, and was so enamored of its shape he created octagonal outhouses to go with it. It was not really until the 1890's that these styles of barns really came into vogue, although they had a weakness that could easily make them structurally unsound. If they were built without a center silo, they had no main support and were easily damaged by heavy snows or strong winds. In the chapters ahead are examples of successful and unsuccessful ones, as they can be found in every state.

Sweitzer (Swiss) forebay or banked barn in Lancaster area. The only barn from the late 1700's I found, showing no additions.

Sweitzer (Swiss) barn somewhere in the Lancaster area, taken during the summer of 1984 when the picture of the previous page was taken. Only recently did I study the early Sweitzer barns built from 1780 to about 1810 and I chanced across this example from my earlier slides. This barn is the only one I have seen in which the original construction appears to be unaltered. Not the cracked wall on the left side, seemingly bound by metal straps. Notice too, the overhanging forbay, because the skeltonal structure of oak beams is not directly tied to the stone walls on either side, they could and have been replaced with wood framework and pine siding on some of the other barns. Since the land appears to be level, I assumed it is ramped on the backside with the wide barn doors

Lancaster area. Majestic is the only word to describe the farmsteads and farmscapes in the hills of this county. One is reminded of a Grant Wood painting in which every house, every barn has been painstakingly placed

The Amish Farmstead

The Amish farmstead is a refuge, an island out of time, where an Amish family may live out their lives in harmony with nature and their surroundings. No harsh sounds of tractors or combines cut the morning air to disturb bird songs, roosters crowing, and the horses' soft neighing and stamping as they await their daily chores, done in tandem with man, woman and child.

A soft summer breeze brings the smell of flowers which surround the vegetable garden with its ripe strawberries, its climbing green beans and peas, its cabbages, carrots and beets. Nearby, the apple, pears, peach, plum, and cherry trees are beginning to bear fruit. Not far away is the grape arbor that shelters the hand pump that used to be the house's only source of water.

The house, the barn, the sheds, and the chicken house are not just appendages stuck on to a working farm. They are the center of family and community life--here church is held, ice cream socials, funerals, frolics, sings. This is the place where the Amish will labor during a lifetime that may span 90 and more years.

The above farmstead, the Swiss barns, and all photographs thru page 95 were taken in and around the Lancaster area.

Here is how one author described it: "The white fence and gate separate the garden from the world outside. Within the Amishman feels secure. It is his home, his church and his livelihood. Within its protection he is born, he worships, he learns and practices his occupation, he courts his wife, he marries her, his own child is born. Here he will die and in his home the funeral will be held. Finally he will be buried in the family graveyard behind another white fence within sight of the one that has sheltered him all his life."

What it it the rest of America searches for in Amish life? Perhaps it is the sense of peace and oneness with nature that the Amish experience in living with the earth. Perhaps it is being in the security of a society where grandfather, father and son all speak the same language--farming, weather, crops, price of good farmland and good barns. How many of us today must wonder if our jobs will be there tomorrow? We must live in fear that suddenly our job will become unnecessary. How many of us yearn to be in touch with the land and although the work may be hard, we know that it will provide for us thru all our tomorrows if we work hard and husband the land. Not so in the outside world were you can give your life to a company, only to find it lost in a hostile takeover and you are suddenly lost? Perhaps people are coming into Amish areas by the million to gaze again on what they have lost--their farm heritage.

Swiss Barns

The Sweitzer barn built on the side of a hill, preferably with the back or upper area toward the north hill, provided a barn that was sheltered from the worst winter winds. German peoples such as Amish and Mennonites dominated the early settling in Pennsylvania They concentrated early on in the Lancaster, Berks, and Chester County areas of Pennsylvania. They brought with them the skills of building structures utilizing huge logs. In Pennsylvania, those skills were honed by the construction of the Swiss or banked barn that was to be the dominant barn found in American agriculture.

    The barns of the cantilever or overhanging forebay type show a distinct relationship to medieval structures in the uplands of the Rhenish Palatinate and on into the Swiss Alps. If one visits this area today one can see anything from homes, churches, and businesses built with an upper entrance at one height, and the lower story having an overhang as a protection from the weather.

Thus someone walking along a medieval street was under the protection of an overhang, which kept rain off and snow from piling itself heavily against the buildings. It was only logical that what worked for people should work for cattle and horses as well. Today no matter where one travels in farmland America, one sees these banked barns with the overhang. It is the most numerous barn type to be found in the Midwest. Even when built on flat land, it was ramped up on one side to provide direct entrance to the threshing floor (or machine shed floor as it is often now called). From there the hay could be unloaded into the haymows on either side.

Unlike the Quakers or Welsh who set up close to towns, the Amish Germans constantly struck out for new territory, stringing their barns and farmsteads along Indian trails. Soon they flowed out of Pennsylvania and into the hills of Maryland and Virginia.

John Heyl in Pennsylvania German Barns, comments: "The skeleton of heavy but expertly trimmed, fitted, and pegged timbers become the functional support and determines the very shape of the great barns...." It is possible to view these skeleton structures in the barn raising photographs to be seen throughout this book. More and more Amish, though, are chosing frame structures to be built more simply with hammer and nails and without the need of at least 100 men to raise the huge and extremely heavy oak timbered frames. I have two hand hewn barn timbers that were 40 feet long in my house and it took many jacks and many days to raise this to an eight foot room height, but the Amish assemble an oak framed structure with some beams raised to a height of sixteen feet and do this in a single day. The one thing this requires is tremendous manpower. All the earlier settlements like Lancaster or Holmes County, Ohio or even Shipshewana, Indiana have the population to support this kind of building... In Minnesota, Missouri and newer areas, it is simpler to build framed barns.

Again from Heyl: "The framing of all these barns was customarily shaped of hardwood timbers, selected for a particular hardness of strength. Hence oak was usually used for the heaviest framing timbers; pegs were often of oak or hickory, sometimes of clearer grained woods such as maple or ash; small membering is occasionally pine or spruce, but the heavy flooring is most usually of sturdy oak planking. The wood siding, both of the exterior and the boarding set tightly about the granary space and defining the stalls, was usually of pine. Sawn timbers first appear as the smaller members of the framing in barns of the early nineteenth century, but the heavy, pegged timbers of the structure, formed by adze and axe were still being set up after the Civil War period. By that time, the overall size of these barns was so exactly established that in later years, when one structure burned, another, no longer in use, could be found, dismantled and re-erected within the fire-scoured but intact masonry walls of the other."

Heyl comments that many of the earlier barns built of stone were built over time, but inside one always found the framework of oak. What seems logical is that a man in pioneer America might set a special day when hundreds of friends and neighbors gathered to put up the skeleton framework of oak, in the space of a day. Then if necessary, he could carry on alone, setting the stone for the walls of a stone barn. I received word from an Amishman whose great great grandfather had lived on the farm that it took two years to lay up the stone for the ends and sides of the barn, yet the basic barn framework was probably assembled by a group of men in a single day.

As barn building rolled across the country, it adopted different materials for the foundation, from sandstone, limestone and shale and finally to brick, but what remained an almost universal was the red color. Again, Heyl explains that " ruddiness of the Pennsylvania German barn is almost proverbial." From early colonial times, deep-toned red iron oxide paint was used to dress the board siding of these buildings. Around window, door and other areas white lead paint or lime whitewash was used to outline simple forms or to transform rectangles into arched or triangular forms.

So it seems that America owes the German people if not specifically the Amish, for the design, shape and color of the American barn. Yet there seems strong evidence as reported by an early diarist that it was the German Amish farms that were admired for their neatness as well as their productivity.

86

The Amish, too, were also in the vanguard of settlers. When the frontiers were opened, the Amish were not far behind. By 1756, they had moved into Somerset County, Pennsylvania, well westward from Lancaster. By 1796, they moved into Big Valley in Mifflin County, Pennsylvania; by 1808 they were moving into Holmes County, Ohio. By the early 1820's, they were settling in Elkhart Coounty, Indiana, and by 1838, the first Amish had reached Iowa, just after it ceased to be Indian territory. By the 1840's they had settled in Washington and Johnson Counties in Iowa. In less than 100 years they had swept across Americas farmlands setting up the Amish farmstead which was duplicated again and again as others copied their highly successful lifestyle and farming techniques.

Following Route 283 out of Harrisburg I found the following barns along the route into the Lancaster area.

Barns of Pennsylvania

If Philadelphia is the cradle of American independence, then Lancaster is surely the cradle of America farm heritage.

By the early 1700s a trickle of German, Amish and others began coming to Pennsylvania at the invitation of William Penn. The trickle became a flood and by 1790.

This was the year Philadelphia became America's capitol, and George Washington signed a bill acknowledging the national debt. By this time the Lancaster area had become the breadbasket of America.

The Amish and Mennonite farmers were acknowledged as the most modern farmers of their day.. Full yone third of Pennsylvania was German and very likely a higher percentage in the Lancaster area. It was said that unless one knew German one had to go with an intepreter

Beautiful stone barns were still being built in 1803, when Ohio became a state. And by then the Amish had spread across Pennsylvania, moving into Somserset County around 1756 and into Mifflin County in the 1790s.

Today Pennsylvania remains the state with the largest number of individual settlements. Lancaster is well known but their are other Amish groups living in 44 of the states counties.

Today if one were to go mining for barns the way Americans once searched for gold, the Lancaster area would be the mother lode of farm/barn history.

In Berks, Chester and Lancaster Counties following no particular roads or pattern one can discover hundreds of examples of barns built two hundred and more years ago.

With the idea of discovering the earliest barn I could find in a single day in the Lancaster area, the first barn I stopped at east and slightly north of the Lancaster motel strip was built in 1803, according to its cornerstone. By noon time I had discovered a barn built in 1790, Near Gap this barn was one of two I found that day with a definite date. The Amishman cut short his lunch to crawl around inside with me and removed a number of bails of hay to expose and area next to the wall; where the date of building had been written in the slack lime cement used to seal the rock walls. Here also were the most complete set of Roman numerials which were carved on each main post and it's accompanying braces. This may have been done because they were done at another site then hauled, probably in winter, by sled to the site and assembled the following spring. This forebay type barn is a forerunner of the all wood barns, banked barns that can be found across the midwest. These barns are most numerous in and near the Amish migration path. The stone barn was built around the inner skeleton of oak barn beams and could be done more slowly after the initial frame and roof were constructed.

A very simple tour off Route 30 to 340 East, this will take you thru Bird-in-Hand, Intercourse, and on to White Horse then take 897 south to Gap then west on 741 to Strasburg. At Strasburg if you take 896 south you will come to the rim of a valley that affords a spectacular view from the rim of the valley, turning around you can follow 896 north until it rejoins thirty. This is a simple easy route to follow which will take you in less than a day past some of the earliest barns to be found in America--again alive and well and functioning on Amish farms.

You can explore for days without running out of barns but approaching Lancaster (the town) by coming east and southward on Route 283, a double lane interstate type highway gives you a panoramic view of the progress of the American barn. From the point were I joined this road just off the Pennsylvania turnpike I found that I found barn style after barn style in sight from either side of the road. There are numerous exits that allow you to get off and examine individual barns or areas but the most historic ones are to be found in the tour above.

I must warn the barn explorer that the Amish might easily welcome an individual traveler, or at least allow him to explore the barn, but I have not identified exact locations for individual barns ,because I did not want the Amish who helped me explore their barn to suddenly be overrun with bus tours. There are a number of show farms in the area which demonstrate how the Amish live and use their barns. Please explore the side roads for further discovers. I have simply given the main tourist trails as a jumping off point.

Today a farm is many things, a place of corn and beans, a place of confinement and specialization, along with mechanization and expansicn; where hogs are processed in such an efficient way that between birth and slaughter they may never be allowed in the open spaces.

In the past, a farm could be many things--a tree shaded stone cottage in New England with a modest barn; a soddy on the rolling priairie of Nebraska; a neat cluster of farming buildings tucked in the lee of a hill in Minnesota, Wisconsin, or Michigan, to protect it from harsh winter winds.

All these images have been shattered and replaced by the pole building, barns filled with heavy machinery, and an absence of life. One farmer in the seat of a bemoth combine can handle two thousand acres of corn or wheat. What is left is an empty landscape filled with row upon row of crop. Yet all is not lost when one looks to the Amish farm. Children scurry to gather eggs, to water the horse and young Amish girls looking the perfect picture of America's farm past herd the cows in and begin their milking chores.

BIG VALLEY

One of the most strikely spectacular settlements of Amish is big valley. Next to Lancaster it is my favorite Amish area in Pennsylvania, different from Lancaster in that it is home to white buggy Amish, yellow buggy Amish, and black buggy Amish. Look for State College, Pennsylvania off Interstate 80 and you have the approximate area. Look for Lewisburg further east and you have the makings of a one day driving tour thru and expanding Amish area. To take the long version begin at Lewisburg and come westward on Route 45, Amish have moved all along this route and you will see them working with horses and mules in front of their large barns set further back up the hill. Just before Old Fort on the right you will encounter the red round barn I featured in the color section, past State Colleg (south of the city) 45 joins 26 which takes you to McAlevy's Fort , then take 305 south into Belleville, follow 655 east until you join 322 running north and south, this route will carry you back up to 80 or down to the Pennsylvania turnpike at Harrisburg, where you can join the road 283 to Lancaster that I wrote of earlier.

Page 96 thru page 99 are photographs taken from Big Valley, Somerset County, Pennsylvania and along Route 45 headed east toward Lewisburg.

The American Barn as a Social Center

Many people today have forgotten that the barn in American farm life was much more than shelter for animals and feed. It was a production factory that helped to feed millions and a social center for Sunday preaching by circuit riding ministers, and a place for barn dances. The difference between the modern farm and the Amish farm is that the Amish barn is still very much used for church--ordinary Sunday church and special occasions such as funeral, wedding, and baptisms. And when parents go off to visit cousins in Kansas, it can quickly become the site of a barn dance as fast as the word can be whispered across the county, which in most Amish areas takes about a day.

The barn is work center and social center on harvest days. Friends and neighbors may gather to bring in the hay and compete to see who is the fastest or who can bring in the most hay, corn, or oats.

If one were to put aside the biggest myth of Amish life--that they are a dour people who don't know much of fun and enjoyment of life, and look back at the farm life of the past, they would remember that one learned to make one's fun as one went along. Work and play were

97

interrelated. The hours of putting up hay, the hours and hour of sweating in the sun beside the horses provided a clean smelling floor for Sunday church and the singings and barn dances the Amish young people had in the evenings.

Both the entertainment and the work revolved around the barn. It was here that the items of machinery were stored, and milk cows had their place. For children the barn was often a chore, mucking out the stalls, but later there would be time to examine the kittens a mother cat had so carefully hid in the hay loft.

Many of the chores of life around a non-mechanized farm require the cooperation of many, yet many working in cooperation make it less of a chore.

Above photo and top of the next page taken around New Wilmington, Pa., near the Pennsylvania/Ohio line.

New Wilmington, Pa.

A unique group of Amish is to be found not far off Interstate 80 on the Pennsylvania, Ohio line. I like to call them the orange buggy Amish, since most of their buggies look a burnt orange color to me. This unique group came from Big Valley and were apparently once yellow buggy Amish. They have a unique approach to current barn building, gambrelled roofed barns are considered too fancy and may not be built. Only plan salt box type-roofs must be used on all barns.

This area is close to Youngstown, were it is reached by driving east on Route 208 out of that city, or by taking 158 south off Interstate 80. Most of the side roads out of New Wilmington have a variety of barns.

Right: In Conewango Valley, New York. The earliest of 12 Amish settlements that can now be found in that state.
Below: Smickesburg, Pa. Fifth largest Amish settlement. This group returned from Ohio to form this colony.

Barns of Ohio

Terms like "fantastic" and "tremendous" come to mind when speaking of this main Amish area of Ohio. I call it Ohio's Amish Heartland, and it is the title of a current book that I am at work on. This area of Holmes, Stark, Tuscarawas, and Wayne Counties with Amish spilling over into Knox and Coschocton Counties is chock-full of some of the most beautiful barns and some of the best scenery to be found anywhere outside Lancaster itself.

This was the first major settlement outside of Pennsylvania and was as close as they could come to another Lancaster or another Rhine Valley of German.barns. Here and in Lancaster, the modest barns and sheds turned massive. Josie Miller has built a barn near Danville, Ohio nearly 140 feet long, just a little shy of half a football field in length.

This is the largest concentration of Amish in the world. Lancaster may have the name, but this area has by far the largest concentration of Amish in one area. Here barn raisings are going on all the time. Amid the twist and turns of these hills, it is not at all unusual to see 300 or more Amish volunteers working on the building of a new pegged barn.

This Amish area is home to Joe T (Josie) Miller and Levi Erb who, like the barn builders of old, could examine the land, listen to what the farmer wanted, assess the nearby stand of timber for lumber. They then take a piece of plan brown wrapping paper and create a drawing of what the finished project should look like and command an army of Amish volunteers to build it in a single day.

From "Mennonite Weekly", I came across the story of Clyde and Ruth Penrod of Holmes County. For those who don't recognize the name as non-Amish, they are not Amish. When a tornado destroyed their barn it was only five minutes after the twister passed that people in buggies started coming down her lane. Said Ruth Penrod, "Thank God for those people."

The Amish took charge and by the second day, Levi Erb had a crew in the Penrod's woodlot cutting the oak trees. They hauled these to the sawmill and brought back the heavy timbers for the rafters.

Opposite page: This 1838 stone barn is on Route 241 four miles north of Millersburg. The following pictures, thru page 106 are details of this and a nearby stone barn built in the same time period.

When all was ready, nearly 250 men showed up to raise the barn and by the end of the day, a 36 wide by 66 foot barn stood where nothing had stood that morning.

Offering to pay was out of the question. They did this for their neighbor as a gesture of friendship and Christian action, knowing that he could not afford to have a new one built. Today, one more monument stands in Ohio's Amish country as a tribute to Amish faith.

When autumn wings its way into this region, usually sometime between October and November, the barns are wonderful spots of red and white among the multihues of the varied trees to be found around this heavily wooded and extremely hilly area. When the first settlers came to Ohio around 1810, they found a region that was densely forested. The first settlers built and burnt and ate from trees. Everything from soap to nuts was provided by the trees. No matter how much was taken away there remained an ample supply of trees to create one of the most picturesque environments in which to view the American barn. Here also are the huge barns as found in Lancaster. Here too, the huge crops are harvested and as the ecological way of Amish farming increased the yields, additions and sheds were added on. Like the Amish home of wood frame, which could be added to as the family expanded, so too was the barn added onto many times as the family prospered

This is one region in which it is difficult to give one tour that would encompass the area and so I will break it down into a number of possibilities for a days drive through barn country.

The first and simplest drive begins on Interstate 77 at Dover, below Canton. Travel westward on 39 through Sugarcreek, and Walnut Creek, and into Berlin. Turn right at the Berlin House Restaurant and head north on 62, back up through Winesburg, and Wilmot, and upward until you reach 30. Then go east back to Interstate 77.

A variation on this is to turn right at Walnut Creek and follow 515 past the Amish Home. It has a barn and home that may be visited along with offering buggy rides. Along this area not far from the home, is the farm where Alma Hershberger attended the barn dance.

Another possibility is to continue on Route 62 &39, past Berlin, and turn left at Route 557 to Charm. On the

far side of Charm you may turn right and wind up on a high ridge overlooking the Doughy Valley, with many a beautiful barn and landscape below.

Continuing to follow Route 62 & 39 west, you can go on to Millersburg. At Millersburg, turn north on Route 83 and follow it into Wooster or onto Highway 30.

In every shop and at the rest stop along Interstate 77, you will find a collection of maps put out by Homes, Wayne, Counties etc. They have excellent little maps about how to get to places along lesser traveled roads. There is so much to see and do in the region. There are so many barns and farms that it would be hard to do it all in a single day. Until recently a traveler had to stay in Dover or New Philadelphia if you wanted a motel room, but more and more motels and campgrounds have been opened up along Route 39. The most unusual one is called Paradise Valley Campground down Route 557 south off 39 & 62, just west of Berlin. The place is up a gravel road off 557 and is not a campground but a series of Amish style log cabins. It is run by an Amish lady. Here you can sleep for the night Amish style in a squeaky brass bed under Amish quilts, read by kerosene lantern, get your water from the outdoor pump, wash in the basin provided, and go to the outhouse for your other problems. It is a chance to experience what it is to live and be Amish. From the high hill you can hear the clomp of horses' hooves on the road below, as buggy after buggy goes by.

A totally different approach starts from Highway 30, either from Wooster or from Canton/Massillon, by going south down 93, 62, 241 or the road into Kidron. Kidron has a Saturday kind of flea market around the town and Country Grocery Shopping Center where you can buy Amish hats and other Amish goods. It is also is the home of Lehman Brothers Hardware where you can still buy a hand operated washing machine, all kinds of wood stoves, and whatever items you want for the barn, from chains to ropes, to pulleys.

One more route is to go from Wooster down 250 South to Strasburg at Interstate 77 or the reverse and back on Highway 30. Each time I visit this area, I try to take a different side road of which there seem to be hundreds.

In preparation for this book, I turned down 241 headed for Millersburg off route 250, and came across a stone barn that reminded me of the Sweitzer or Swiss Barns of Lancaster. This barn was built in 1838 and is only four miles north of Millersburg. I believe it is probably one of the oldest in Ohio at least that is still in use.

Stone barn above from Milersburg, circa 1836. This barn or the one on the previous pages maybe one of the oldest surviving barns, in use, in Ohio.
Right: North off Route 62 near Winesburg. This barn proudly announces 1852 on its front.
Below: One of the few surviving log barns. No one knows the date it was built. This style once was the most dominant barn in this area.

JACOB BARDER, BARN BUILDER

The letter came from Manhattan, Kansas from a lady whose father "built many barns for the Amish in Holmes Co. Ohio. He probably built more barns than anyone else in Ohio, as he spent his entire life building barns in Tuscarawas County." I found his barn near Strasburg. The roof announced "J. G. Barder" in slate on the roof. I also found many slate roofed barns with the name of the owner and the date which I believe could have been his handy work. Here is only the home farm and another barn just north of Mt. Eaton

Many barns along Route 62 north of Berlin seem to date from the 1850's thru the 1860's. This cornerstone reads "J. Hershberger."

Above: This several storied barn was just off Route 551, and its foundation had already been replaced. In the backyard I found the cornerstone "1885".

Below: The Amish Home is a home and barn that can be visited as a tourist attraction. It offers guided tours of both house and barn, and buggy rides. The barn is from the same period as the one above.

Unique describes this ten-sided barn west of Shreve, Ohio. Note the interior view with no center silo or support beam. A huge arrangement of large iron bolts screwed into the dome stabilize what would normally be a weak structure. This barn, built in the 1890's stands as firmly as it did the day it was built.

Just down the road from the ten sided barn is this structure of 100-125 years of age. Its most impressive feature is an oak beam, approximately 12X14 inches thick and 56 feet long. Imagine an oak tree 56 feet tall to the fork!

Eric Sloane from <u>The Age of Barns</u> comments: "As the European was a connoisseur of wine and perfume, the American took to the aroma of good wood."

    In the past every child knew as much about wood as the child today know aluminum from steel. He knew hardwoods from softwoods, hickory for smoking hams, fragrance of pine boughs, for beds and to add a smell to the house, the sap of the maple for the syrup all America still loves--wood was good for building and burning, for soap and for turpentine. It was made into barrels and bowls and wooden spoons--butterchurns, bowls and boxes. There was even a word to discribe products made of wood "treenware".

    The Amish are perhaps the last people in America that goes to the tree for wood and not the lumberyard. In the begging of the book is the story of an Amish barn raising in which 125 trees were logged to provide the 25,000 board feet of lumber needed to build a new barn. Today Amish woods and Amish orchards provide the family with everthing from soap to nuts.

Barn raising near Millersburg, Ohio. Photos by Bruce Glick.

Middlefield/Mesopotamia

Gauga and Trumbull Counties have a rather large Amish population, but it is an area that until recently has had little tourism. What it does have is Amish factory workers, both male and female. Cleveland businesses have discovered that setting up a factory in an Amish area gives them more dependable workers. Route 87 straight east out of Cleveland takes you into the heart of this country, with Middlefield as the center, and Burton and Mesopotamia to either side. Anywhere along 87 are some nice barns, but past Middlefield any road you take south and will go past a succession of Amish farms. This is a nice day's drive if you are visiting in Cleveland and is shorter but less rewarding than going 75 miles or more south into Ohio's Amish heartland.of Holmes County.

Barns of Maryland and Delaware

There are two areas of Amish settlement in Maryland. Oakland, in Garrett County in the far southwestern corner of the state contains one of the most unique Amish groups in America. Some thrilling views of barns both Amish and non-Amish can be seen there. This group has adopted the tractor and a fixed church, which is a plain white building. They go to church in horse and buggies on Sunday, otherwise they drive to town in tractors with little trailers pulled behind for passengers.

Going south from downtown Washington D.C. on Highway 5 will bring you to Charles and St Mary's Counties. From the town of St Charles on you will see grey Amish buggies and signs of Amish life along both sides of the road. This group moved into the area in the early 1940's, the only grey buggy group I know of outside the Lancaster area. They mainly truck farm and I have seen Amish women and children selling bundles of flowers along the main roads near their home. There is a Saturday farmers market and flea market near St. Charles on Route 5.

Near Dover, Delaware

Opposite page: South of Oakland, Maryland, on the southwestern tip of the state. I came across this spectacular view from a hill in the heart of the Amish area.

Dover, Delaware has a fairly large group of Amish living just outside this small state capitol. This follows the Amish desire to be just outside small towns rather than be near big cities. Dover is so small that it could be lost inside many a small Iowa town that the Amish live near. In many respects the farming here is not much different from other areas and the barns again are similar. If you are in the area, they are located west of town, but ask almost anyone in the area for specific directions.

Menno-hof Museum

A striking example of the Amish-Mennonite farm is Menno-hof, north off Highway 20 East, headed into Shipshewana. It was built to look like a picturesque example of the farms that dot Mid-America. It is actually a museum of Mennonite, Amish, and Hutterite heritage. Inside the shell that looks like a farm is a museum that combines slide shows with touch and feel recreations of a prison torture chamber, and interior of a boat resembling ones that brought them to America, and rooms like those found in medieval Europe.

Photos by Mike D. Hostetler

Menno-hof, as the brochure says, combines "Menno" (Menno Simons was the leader who brought stability to the early Anabaptists) with "Hof", the German word for farmstead. From early martyrdom and forced migration from Europe, they found a new life in America, inhabiting buildings like this one, designed to resemble a typical northern Indiana farmstead.

Once within the door, the visitor is treated to over 400 years of Mennonite-Amish history , from its lowly beginnings in Switzerland and Germany, to its state today, with over 750,000 members scattered in 57 countries. In February, 1527, the Anabaptists declared themselves a free church; the first church in over a thousand years to separate themselves from the state. "This stand opened the floodgates to one of the longest and bloodiest persecutions in the history of the church, a campaign carried out by both Catholic and Protestant authorities."

Mennonites and Amish settlers began arriving in northern Indiana by the early 1840's and today, 60 Amish church district and 160 Mennonite congregations are located within an hour's drive from Shipshewana.

The idea for the center began with Harvey Chupp, a Mennonite pastor who worked parttime in the Yoder Hardware Store at Shipshewana. A flea market was begun in 1972, and by the 1980's, was bringing a flood of visitors into the area. Many of them were curious, but totally without knowledge of the Amish and Mennonites. Harvey Chupp began a slide-lecture at a local; church and it finally grew into Menno-hof. an Amish crew and 200 volunteers erected the structure in six days in 1986, but it took two more years to complete the historical recreations inside. It opened to the public May 2, 1988.

Brown Swiss Barn

In 1887 a young Indiana farmer, Menno S. Yoder, wrote for a book about building with concrete. It was not until 20 years and much study that he built a 12 sided concrete barn, completed in 1908. Eric Sloane commented in his barn book of discovering a 12 sided barn, but until I began the final phases of this book, I had seen no more than an eight sided one. Then I found a highly unusual 10 sided one in Ohio.

When he was finally ready to build his concrete barn, he bicycled 40 miles over dirt roads to South Bend to observe the construction of concrete buildings. He and his five eldest sons, (he had ten children) hauled gravel from five miles away for two winters prior to beginning, because he wanted only the best material. He collected buggy axles to re-enforce the windows and rails for the larger doors. The walls are over a foot thick at the base and narrow to six inches at the top. A steel tractor wheel was used in the center to bolt rafters to, and farm fence was used throughout to fashion reinforced concrete.

The barn is located on County Road 16, just outside Shipshewana, headed west toward Middlebury. It seems to have changed little from the time when it was built, a tribute to a thrifty young farmer who wanted a barn you didn't have to paint and required almost no maintenance.

Barn raising, framed barn, west of Highway 6 at Nappanee

Barns of Indiana

The Amish were always in the forefront of farm migration across the county. By the early 1840's they had settled in what was to become Elkhart and LaGrange Counties. Another group had settled in Washington and Johnson Counties of Iowa at almost the same time.

From the beginning they sought isolation from the world and from their close brethren, the Mennonites, from whom they descended. They believe the larger group would influence their much smaller, group broken off from the main body by Jacob Ammann, a former Mennonite minister. Although both were mostly farmers, in the beginning, the Mennonite religion quickly spread to city dwellers and to professionals and tradesmen. The Amish fled westward seeking isolation from all and separation from the Mennonites.

It seems ironic that there are now 160 Mennonite churches in comparison to 60 Amish church districts within one hours drive of Shipshewana. This is the same story in every Amish area in the country and most of the Mennonites are descendants of Amish. A liberalizing element in the Amish church created Amish-Mennonite Churches, which in turn finally dropped the Amish from the front and became simply Mennonite Churches.

Dig into any Mennonite farmer's past and you will probably find Amish grandparents. Thus the very thing the Amish sought to avoid, a close relationship with the Mennonites, they themselves have fostered.

The building of Menno-hof demonstrated the co-operation of both groups in building, barn raising fashion, a museum of their mutual heritage. Everywhere I have traveled people are curious about the Mennonites and Amish and woefully ignorant. What better place to have a museum than nearly across the road from a flea market that now attracts up to 100,000 people on Tuesdays and Thursdays of each week during the summer.

For the barn buff, the nearly 1400 dealers and six auctions offer a unique opportunity to look for barn and farm related antiques and collectibles.

When I began photographing and studying the Amish more than 25 years ago, the entire U.S. Amish population was said to be only around 25,000. Today it is said to be over 100,000, with the number expected to double in the next 20 years.

When I first visited Shipshe it was a dying little town with nothing much left open downtown. Today, house after house is being taken over to be turned into shops and it is overrun with people.

No other towns except Berlin and Walnut Creek, Ohio and Bird-in-Hand and Intercourse, Pennsylvania have more tourists coming to examine Amish life. Menno-hof makes an ideal jumping off point for the exploration of barns in this area. Leaving Menno-hof at Shipshe on Route 5, just north of U.S 20, the traveler has many choices. Going north on 5 will take one past many Amish farms. A left turn at the main intersection of town will go past the Brown Swiss Barn and down a road leading to Middlebury. At Middlebury a left turn south down Route 13 leads past a beautiful red dairy barn of the banked type built around 1865.

U.S. 20 goes past Essenhaus Restaurant which began life as a barn and has now become one of the most famous restaurants serving Amish style cooking in the country.

Route 5 North to 120, 120 west to 13, 13 south to U.S. 20, then east will square an area in which there are hundreds of beautiful barns to be seen, not to mention all the horsefarming activities that are constantly going on in the fields surrounding these farmsteads.

I examined a newly built peg barn shop at the gift shop in downtown Shipshewana. It was being built by Big Dave Bontrager just as I was completing this book and should by now be open to the public.

Following Route 5 south to U.S. Highway 6, then west until coming to Nappanee, the traveler will be in another Amish area in the lower edge of Elkhart County. On Highway 6 is Amish Acres, which contains an authentic Amish home, and a recreated barn gift shop, since the original barn burned down. South of the highway are a lot of Amish homes and northward on Route 19 near Wakarusa live the Old Order Mennonites, who also use the horse and buggy and participate in barn raisings. Just west of Amish Acres on Highway 6 is the newly created Dutch Village, a huge antique complex which also features antique dealers and antique auctions, for anyone looking for barn memorabilia.

Two more areas of Indiana deserve attention for their selection of barns and Amish life. Grabill, Indiana is in Allen County east and north of Fort Wayne off Route 37. Grabill is home to a group that does not believe in having tops on their buggies. A drive through the area on Sunday, on roads near the town will allow one to observe Sunday church services taking place in area barns. There is usually quite a group of buggies both up and down the roads as well as on the farm itself.

Berne, Indiana is similar to Grabill in that the Amish are strict and do not have tops on their buggies. This unique group moved in the 1850's directly from the Canton of Bern in Switzerland, to the town of Berne, on Route 27 about 40 miles south of Fort Wayne. I believe they probably founded the town. A unique feature of this group is that it demands that the barns demand that the barns be white only. Any red barns in the area belong to non-Amish farmers.

This defective round barn is on the County Line Rd, south and west of Nappanee, between Marshall and Kosciusko Counties, Built in 1910, a severe windstorm in the late 30's canted it and the farmer rushed out and got old long barn beams to shore up the tilting roof. It had no center silo for a brace. An examination reveals many hand hewn timbers from some earlier barns.

On roads near town, one can observe Sunday church services in a barn, as below. A unique sight is this 100 odd year old barn jacked up on a semi-trailer for removal to another farm, since a new city house was built nearby.

Grabill, Indiana, only a short distance from Fort Wayne up route 37 north and east is a unique settlement. Founded in the 1850's, it is the home of a group that does not believe in having tops on their buggies.

Barns near Berne, Indiana. This very strict community moved directly from Bern, Switzerland. These Amish don't believe in red barns. If they take over a non-Amish farm they must paint the barn white, so if you see a red barn as above, it is not Amish. This is the largest of four direct migrations of Anabaptists from Europe to Indiana.

Barns of Illinois

Illinois is a latecomer among the earlier Amish settlements in America, possibly because the Amish overlooked it for the more hilly wooded lands of Iowa. However, by the 1850's they had discovered what is now know as the Arthur/Arcola area and began to spread across the flat land of Douglas and Moultrie Counties. This remains the only real settlement of the Amish in Illinois, in marked contrast to its sister state Missouri. The excellent barn with the Mail Pouch sign on it was right on 133 East just off Interstate 57, which brings you down to this area from Chicago or Champaign-Urbana. This area is also not far from Springfield, so that those visiting this area might visit Lincoln country as well. Lincoln and his beard and even his way of life would not be out of place among the Amish of this region. And Route 133 West takes your right thru the heart of the Amish area, although I recommend you take a left fork off that route, just west of the town of Arcola and follow that excellent black top road. Plenty of barns are to be

found along it and you may take it west until you see a sign for Arthur. Turn right for Arthur and then return east on 133. On the side road, you will see a sign for Rockhome Gardens, a Mennonite farm that has become the leading tourist attraction of the area. Unlike many such establishments in Amish areas, it is open on Sundays and in addition to exhibits of Amish homes and barns which may be toured, it also has an excellent restaurant.

There are many excellent examples of barns to be found in this area, but when I questioned an Amishmen at a local store, he said that it had been more than 30 years since he can remember a real barn raising in the area. So although this area does have some excellent examples of the earlier pegged barns, all the newer ones tend to be framed ones or even pole buildings.

This is one of the areas most solidly dominated by the Amish and unlike many other areas, you are liable to see from dozens to hundreds of buggies in Arthur during a Saturday morning shopping expedition.

Barns of Missouri

How to make a long story short--Missouri is the fastest growing state in terms of Amish population. New settlements stretch from the top to the bottom of the state. Jamesport is the earliest settlement, from 1953, and may be reached by going west from Kirksville on Highway 6 or west from Interstate 35. But this is only the beginning. The second oldest is north of St Louis on Route 61 at Bowling Green, with an increasing number spreading southward down Route 63, the road that took us through the newest Amish area in Iowa. There are Amish at La Plata, near Anabel in Macon County, Clark in Randolph County. Jumping southward past Jefferson City, below Bagnell Dam and the heart of the Ozarks, Amish can be found around the towns of Seymour and Marshfield, near Springfield, Missouri. Missouri is one of the farm states that experienced a depression and everyone went to Kansas City or St Louis to find work. It is open country for Amish farmers looking for small towns away from large population centers. Besides the Amish areas, Missouri itself has a rural and small town life worth exploring. I have future plans to do a barn book on the state which would include all the Amish areas.

Barns of Kansas

Of all the Midwestern states, Kansas is unique in Amish culture and barns. Alone among Amish areas, it has more dependence on mechanization and cash crops. The Amish farmers living near Yoder and Hutchinson adopted gas-powered combines to harvest wheat when they found that horses died in the hundred degree heat of harvest time. The primary area is north and a little west of Wichita on Route 96. Here the great migration in 1875 of Mennonites from Russia, and Amish from Prussia settled, bringing a tough wheat called Turkey Red. With it they tamed the plains of Kansas and turned it into the nation's breadbasket, just as the Amish had 100 years earlier in Lancaster.

The barns tend to be western style with long sloping roofs and clustered about them, numerous sheds and other buildings. Although they differ from farmers in other areas, giving many more acres to wheat, they still plant oats for their horses. The horses are used not only for the buggies, but to haul the grain wagons to town. Once the wheat is harvested by the large combines, they switch to horsedrawn wagons to haul it into town. Because of their smaller number of horses, large barns for hay storage are not necessary.

More traditional Amish farming methods are employed at Garnett. Follow Interstate 35 south and west of Kansas City, Kansas, then Route 59 south to Garnett.

Barns of Iowa

When the American frontier opened to settlement, the Amish were in the forefront of the farmers. Iowa was officially open to settlement in 1833 and by 1838, the Amish came to Henry County. It was not until the 1840's that they settled into the Washington/Johnson County area near Iowa City, which is right on the edge of Interstate 80.

At the time of settlement, those early Amish would get on a trail cut by twenty yoke of oxen from Dubuque to this area and walk or ride the more than 100 miles to register their claims. Remnants of this early road across thick virgin prairie still exists a little off Highway 1, near the town of Joetown.

Barns of the area which is rolling hills again copy the barns of Lancaster. The banked or forebay type barn predominates, although they can be found as gambreled roof or even octagonal. It is not difficult to find barns that go back one hundred years, but unlike Pennsylvania and Ohio, I have not been able to find cornerstones or carved dates in timbers for identification.

The region resembles all primary settlement areas in that it tended to be among rolling hills. The largest part of the Amish population lives in the English River Valley. According to historical reports, the area had many trees of oak and walnut which could be used to build sound peg barns. Hundreds survive in beautiful condition in this area.

Just before I concluded this book, I took a quick tour of the Amish barn area. Taking a road two miles northeast of Kalona, I went eastward nearly a mile and stopped on a rise of ground near a church and cemetery. By turning in a full circle, I could count within eyesight, twenty clusters of barns. All were in good shape and in good repair. They represented nearly every kind of barn that can be found throughout the North American continent with the exception of round or octagonal. These are getting rare, yet within a few miles of this area this styles of barn exists, also.

As I began, so shall I end. **"The American Barn is Alive and Well and Living in Amish America!**

Kalona Barn Tour

To keep things simple, I have laid out tours from nearby cities through the Amish areas that can be completed in a single day. They are hard to get lost on and if you do get lost, you just might discover some interesting barns.

From Iowa City, get on Highway 1, either by joining it at Iowa City at Interstate 80, or by finding Riverside Drive along the Iowa River and proceeding South and West. Five miles from town on a hill is a major road that forks to the left. A road sign points toward the Welsh church and to the Welsh community of Old Man's Creek. If you will turn left here, immediately past an old iron bridge about a mile down the road, you will see to the left a marvelous example of an octagonal barn. Again this is a banked barn built by a Welshmen, but it is now on the doorstep of Amish America. On either side of the road past this point you will encounter many Amish farms and barns. (see photos) By the time you have passed what is left of the little town of Sharon Center you are in the heart of Iowa's main Amish area. Here you might want to diverge to right or left to explore, but to keep on track always return to the main blacktop. A mile or so out of Sharon Center, it is crossed by another black top, the Cheese Factory Road, named for the cheese factory on the corner of it and Highway 1. You could swing west and head back up to Iowa City, if you did not have much time, but the next four miles straight ahead take you through the heart of Amish barn country. The road will become gravel, but proceed on it until you come to Highway 22, east of Kalona, turn right or west and you will head into Kalona. You may want to take a stop at the Kalona Historical Village on the edge of town. Continuing straight will take you to Highway 1, where you can once again turn north and head back to Iowa City. Highway 1 has some of the most beautiful of Iowa's red barns. Driving this route you will be able to see several hundred barns from several years old to over 100 years old. Keep in mind that there are many more to be found on the sideroads to the right and left, and it is easy to get back on that main trail, Highway 1.

Oelwein, Iowa

Bloomfield/Milton

Oelwein--Independence.

In the 1920's, the Amish moved from the Kalona area to Oelwein and Independence. This was where Alma Hershberger grew up. Alma wrote a piece in this book on Amish barn dances. Her book, Amish Life's Thru a Child's Eyes is listed in the front of this book.

Many fine barns can be found in this area accessible from Interstate 380 North, by leaving it to head north on 150 which takes you through both Oelwein and Independence. South of Oelwein, no more than a mile is 281, which goes west to Fairbank. From here a blacktop goes south, taking you through the Amish area and back down to Independence, following a rough circle. From there you can go east or west of 20. By following 20 west, you will rejoin 380 at Waterloo, or you may simply go back south on 150 to 380 where you exited.

Bloomfield/Milton in Davis and Van Buren Counties

This community began in the last ten years or so. It has grown so fast with immigrants from eastern and northern states that it has almost as many Amish as the Kalona area which existed for almost 150 years. Off the beaten path, this settlement is reached by going south on 63 from Ottumwa, Iowa until you reach Bloomfield, then proceeding east on Route 2. What you are likely to see is older barns redone and fresh ones recently built, of concrete block and tin along with storebought lumber, but an area nevertheless growing in its number of barns.

Outside Kalona. Amish graveyards on hills overlook the farmlands where the Amish lived out their lives. This scene in Iowa is echoed again and again in other Amish areas across the country.

The Farm Today and Yesterday

Today a farm is many things, a place of corn and beans, a place of confinement and specialization, along with mechanization and expansion; where hogs are processed in such an efficient way that between birth and slaughter they may never be allowed in the open spaces.

In the past, a farm could be many things--a tree shaded stone cottage in New England with a modest barn; a soddy on the rolling prairie of Nebraska; a neat cluster of farm buildings tucked in the lee of a hill in Minnesota, Wisconsin, or Michigan, to protect it from harsh winter winds.

All these images have been shattered and replaced by the pole building, barns filled with heavy machinery, and an absence of life. One farmer in the seat of a behemoth combine can handle two thousand acres of corn or wheat. What is left is an empty landscape filled with row upon row of crop. Yet all is not lost when one looks to the Amish farm. Children scurry to gather eggs and to water the horse and young Amish girls looking the perfect picture of America's farm past, herd the cows in and begin their milking chores.

Above and on the following page are barns from old Highway 1, the road that runs thru Sharon Center. Note the Roberts' barn, an octagonal, built by a Welshman.

A Typical Day

Our day started about 5:30 in the morning. We milked from 20 to 40 cows by hand. We had dairy cattle, beef cattle, hogs and sheep. We farmed with horses, except for plowing and harrowing. for that, we had a big steel-wheeled tractor with a three bottom plow.

We chored and milked before we went to high school or grade school, about an hour and a half, morning and evening. We ground feed on Saturday-for the cows. We cut firewood for the school and church on Saturdays too. Both had wood furnaces. We walked to school or rode horseback. We all came up thru the local one room schools and the local high school at Sharon Center, which our parents helped build.

The Winborn family, Sharon Center, 1930
(Excerpted from <u>Iowa Barns No.1</u>)

Zip Code and P. O. of Amish Communities in U.S.A.

ARKANSAS
72102-McRae
72143-Searcy

DELAWARE
19938-Clayton
19901-Dover
19953-Hartley
19955-Kenton
19934-Wyoming

FLORIDA
33578-Pinecraft
33577-Sarasota

GEORGIA
31063-Montezuma

ILLINOIS
61911-Arthur
61910-Arcola
61931-Humboldt
61937-Lovington
61951-Sullivan
61953-Tuscola

INDIANA
46911-Amboy
46711-Berne
46506-Bremen
46507-Bristol
47326-Bryant
46800-Fort Wayne
46737-Fremont
46740-Geneva
46526-Goshen
46741-Grabill
46936-Greentown
46742-Hamilton
46746-Howe
46747-Hudson
46901-Kokomo
46761-LaGrange
46538-Leesburg
46767-Ligonier
47553-Loogootee
46540-Middlebury
46542-Milford
46150-Milroy
46543-Millersburg
46772-Monroe
47558-Montgomery
46550-Nappanee
46774-New Haven
47562-Oden
47452-Orleans
47454-Paoli
46779-Pleasant Lake
47371-Portland
46173-Rushville
47167-Salem
46565-Shipshewana
46787-S. Whitley
46788-Spencerville
46571-Topeka
46795-Wolcottville
56797-Woodburn

IOWA
52537-Bloomfield
52552-Drakesville
50629-Fairbank
50641-Hazleton
52240-Iowa City
50644-Independence
52247-Kalona
50144-Leon
50455-McIntire
52570-Milton
50466-Riceville
52327-Riverside

KANSAS
66032-Garnett
67543-Haven
67501-Hutchinson
66091-Welda
67585-Yoder

KENTUCKY
42728-Columbia
42217-Crofton
42528-Dunnville
42234-Guthrie
42064-Marion
42528-Nunnville
42160-Park City
42286-Trenton

MARYLAND
20622-Charlotte Hall
21645-Kennedyville
20659-Mechanicsville
21550-Oakland
21678-Worton

MICHIGAN
49232-Camden
49032-Centreville
48617-Clare
49040-Colon
49310-Blanchard
49028-Bronson
49030-Burr Oak
48624-Gladwin
48838-Greenville
49245-Homer
49431-Ludington
49072-Mendon
48647-Mio
49082-Quincy
49274-Reading
48878-Rosebush
49454-Scottville
48888-Stanton
49091-Sturgis

MINNESOTA
56437-Bertha
56438-Browerville
55922-Canton
55923-Chatfield
56453-Hewitt
55939-Harmony
55965-Preston
55972-St. Charles
55979-Utica
56182-Wadena

MISSOURI
63431-Anabell
63334-Bowling Green
65243-Clark
63435-Canton
63339-Curryville
65034-Fortuna
64648-Jamesport
63549-LaPlata
65050-Latham
63552-Macon
65263-Madison
65706-Marshfield
63463-Philadelphia
63454-Maywood
65746-Seymour
65360-Windsor

MONTANA
59930-Rexford

NEW YORK
14711-Belfast
14719-Cattaraugus
14433-Clyde
14724-Clymer
14726-Conewango
14728-Dewittville
14739-Friendship
13654-Heuvelton
13416-Newport
13667-Norfolk

OHIO
44606-Apple Creek
44003-Andover
44805-Ashland
43804-Baltic
43310-Belle Center
44610-Berlin
44611-Big Prairie
44021-Burton
44822-Butler
44617-Charm
44618-Dalton
43014-Danville
44624-Dundee
44627-Fredericksburg
43019-Fredericktown
43824-Fresno
44231-Garrettsville
44837-Greenwich
44632-Hartville
43526-Hicksville
44633-Holmesville
44235-Homerville
43028-Howard
44046-Huntsburg
43324-Huntsville
43326-Kenton
43332-LaRue
44638-Lakeville
43755-Lore City
44842-Loudonville
44062-Middlefield
44654-Millersburg
44657-Minerva
43340-Mount Victory
44662-Navarre
43832-Newcomerstown
44663-New Phila.
44450-N.Bloomfield
44667-Orwell
44076-Orwell
43064-Plain City
44866-Polk
43837-Port Washington
43773-Quaker City
43345-Ridgeway
43347-Rushsylvania
43778-Salesville
43988-Scio
44878-Shiloh
44676-Shreve
44681-Sugar Creek
44880-Sullivan
44685-Uniontown
43080-Utica
45786-Waterford
44491-W. Farmington
44287-West Salem
45693-West Union
44689-Wilmot
45697-Winchester

OKLAHOMA
74337-Chouteau
74535-Clarita
74538-Coalgate
74036-Inola
73669-Thomas

PENNSYLVANIA
16820-Aaronsburg
17302-Airville
17002-Allensville
17810-Allenwood
16111-Atlantic
17812-Beaver Springs
16823-Bellefonte
17004-Belleville
17505-Bird-in-Hand
17815-Bloomsburg
16404-Centerville
17509-Christiana
16406-Conneautville
17821-Danville
16222-Dayton
17220-Dry Run
17314-Delta
19520-Elverson
16120-Enon Valley
16124-Fredonia
17527-Gap
17529-Gordonville
16327-Guys Mills
16038-Harrisville
17531-Hessdale
17532-Holtwood
15747-Home
19344-Honey Brook
16841-Howard
16133-Jackson Center
17535-Kinzers
17536-Kirkwood
17600-Lancaster
17042-Lebanon
17540-Leola
18829-LeRaysville
17543-Lititz
17340-Littlestown
17747-Loganton
17049-McAlisterville
17841-McClure
17759-Marion Center
16137-Mercer
15552-Meyersdale
17844-Mifflinburg
17059-Mifflintown
17751-Mill Hall
17063-Milroy
16854-Millheim
17847-Milton
17752-Montgomery
19543-Morgantown
17853-Mt Pleasant
17067-Myerstown
17555-Narvon
17240-Newburg
17557-New Holland
17073-Newmanstown
17560-New Providence
16142-New Wilmington
16362-Nottingham
17563-Peach Bottom
17562-Paradise
17864-Port Trevorton
15767-Punxsutaweny
17566-Quarryville
16872-Rebersburg
17084-Reedsville
15771-Rochester Mills
17572-Ronks
15558-Salisbury
17870-Selinsgrove
17073-Sheridan
17257-Shippensburg
17577-Soudersburg
16434-Spartansburg
16256-Smicksburg
17262-Spring Run
15562-Springs
16153-Stoneboro
17579-Strasburg
16350-Sugar Grove
16353-Tionesta
16360-Townville
15866-Troutville
17772-Turbatville
16438-Union City
16156-Volant
18851-Warren Center
17777-Watsontown
17889-Winfield
16882-Woodward
18853-Wyalusing

TENNESSEE
38224-Cottage Grove
38456-Ethridge
38341-Holladay
38344-Huntingdon
38464-Lawrenceburg
38384-Summertown

VIRGINIA
22709-Aroda
22019-Catlett
23452-Lynnhaven
22956-Mission Home
23456-Prince Anne
22738-Rochelle
24477-Stuarts Draft
22744-Twymans Mill

WISCONSIN
54406-Amherst
54722-Augusta
54616-Blair
53520-Brodhead
53923-Cambria
54619-Cashton
54728-Chetek
54422-Curtiss
53926-Dalton
54436-Granton
54437-Greenwood
54451-Medford
53574-New Glarus
54648-Norwalk
54651-Ontario
54460-Owen
53954-Pardeeville
53901-Portage
54958-Princeton
54479-Spences
54667-Westby
54670-Wilton

Grandpa's Barn

Grandpa said:
   "I'll build her straight,
   And I'll build her well,
   And I'll build her so
   She lasts a spell."

Then he sat down
when the roof was on,
And read a bit
from the Gospel of John.

He made her straight
in the morning sun.
Never quit,
'til the place was done.

When I stand there,
in the old barnyard
And think how Grandpa
worked so hard,

I'm grateful for a heritage
up from the sod
With a place for work
and a place for God.

My heart says:
   Build it straight,
   Build it well,
   Build it so
   It will last a spell.

Robert K. Perrin